THE
Blue
Zones

THE

Blue
Zones

LESSONS FOR LIVING LONGER
FROM THE PEOPLE
WHO'VE LIVED THE LONGEST

DAN BUETTNER

NATIONAL GEOGRAPHIC

WASHINGTON, D.C.

For Roger and Dolly

Published by the National Geographic Society
Copyright © 2008 Dan Buettner
All rights reserved. Reproduction of the whole or any part of the contents without
written permission from the publisher is prohibited.
ISBN 978-1-4262-0274-2
ISBN 978-1-4262-0400-5 (trade paperback)
First paperback printing 2009
Printed in the U.S.A.

The Library of Congress has cataloged the hardcover edition as follows:

Buettner, Dan.
  The blue zones : lessons for living longer from the people who've lived
the longest / by Dan Buettner.
    p. cm.
  Includes bibliographical references and index.
  ISBN 978-1-4262-0274-2
  1. Longevity. 2. Medical geography. I. Title.
RA776.75.B84 2008
613.2--dc22

2007044375

Founded in 1888, the National Geographic Society is one of the largest nonprofit scientific and educational organizations in the world. It reaches more than 285 million people worldwide each month
through its official journal, NATIONAL GEOGRAPHIC, and its four other magazines; the National
Geographic Channel; television documentaries; radio programs; films; books; videos and DVDs;
maps; and interactive media. National Geographic has funded more than 8,000 scientific research
projects and supports an education program combating geographic illiteracy.

For more information, please call 1-800-NGS LINE (647-5463)
or write to the following address:

National Geographic Society
1145 17th Street N.W.
Washington, D.C. 20036-4688 U.S.A.

Visit us online at www.nationalgeographic.com

For information about special discounts for bulk purchases, please contact
National Geographic Books Special Sales: ngspecsales@ngs.org

For rights or permissions inquiries, please contact National Geographic Books Subsidiary Rights:
ngbookrights@ngs.org

Interior design: Cameron Zotter

10/RRDC/5

A MESSAGE TO THE READER

This publication contains the opinions and ideas of its author. It is intended to provide helpful and informative material on the subjects addressd in the publication. It is sold with the understanding that the authors and publisher are not engaged in rendering medical, health, or any other kind of personal professional services in the book. The reader should consult his or her medical, health, or other competent professional before adopting any of the suggestions in this book or drawing inferences from it.

The authors and publisher specifically disclaim all responsibility for any liability, loss, or risk, personal or otherwise, which is incurred as a consequence, directly or indirectly, of the use and application of any of the contents in this book.

Contents

Acknowledgments

WITHOUT THE UNIVERSITY OF MINNESOTA'S Dr. Robert Kane, who endorsed and helped shape the Blue Zones premise, this book would have never materialized. He and his colleagues from the National Institute on Aging, Dr. Jack Guralnick, Dr. Luigi Ferrucci and Dr. Paul Costas; Dr. Thomas Perls from the New England Centenarian Study; Dr. Greg Plotnikoff, Medical Director of Allina's Institute for Health and Healing; University of Lovain's Dr. Michel Poulain and University of Illinois, Chicago's Dr. S. Jay Olshansky would spend countless hours sharing expertise, identifying locations, developing methodologies, and ultimately keeping me on the path of science and off the short cuts of conjecture and hyperbole. I cannot thank them enough.

Of the many experts around the world who contributed to this project, I am especially indebted to Dr. Craig Willcox, Dr. Bradley Willcox, Dr. Mokoto Suzuki of the Okinawa Centenarian study; Dr. Tatsama; Dr. Luca Deiana of Sardinia's AKEA Study and his incandescently brilliant protégé Dr. Gianni Pes; Dr. Paolo Francalacci; Drs. Gary Fraser and Terry Butler of the Adventist Health Study; Dr. Luis Rosero-Bixby of the Central American Population Center; and Dr. Leonardo Mata. They not only lent their expertise but also extended their hospitality and generosity of spirit. Dr. Len Hayflick, Dr. Jack Weatherford, and Dr. Richard Suzman graciously consented to many long interviews. The faculty at the University of Minnesota's School of Public Health, including Dr. Robert Jeffreys, Dr. Tatyana Shamliyan, Dr. Robert W. Jeffery, Dr. John Finnegan, Dr. Cheryl Perry, and especially Dr. Leslie Lytle have been and still are my academic partners.

Many of the experiences on which this book is based reflect a shared effort by the members of Quest Team who have traveled with me to the Blue Zones. Photographer and long-time expedition partner David McLain deserves much of the credit in developing the Blue Zones idea. Nick Buettner, Damian Petrou, Gianluca Colla, Sabriya Rice, Rachel Binns, Sayoko Ogata, Dr. Elizabeth Lopez, Eliza Thomas, Tom Adair, Michael Mintz, Meshach Weber, Thad Dahlberg, Eric Luoma, Joseph Van Harken, and Suzanne Pfeifer all shared their ample talents and endured many long days and nights to bring Blue Zones to life.

This story would have never been told without Peter Miller, my editor at National Geographic. He backed the idea for the original magazine story and guided me through my first drafts of the book. Michelle Harris further improved the book through her thorough fact checking, and Dr. Robert M. Russell's review of our chapters helped keep us on track. Also at National Geographic, I thank Lisa Thomas and Amy Briggs for orchestrating this book; Rebecca Martin for shepherding us through the Expeditions Council grant process; Valerie May and Miki Meek for bringing Blue Zones to life online; and picture editor Susan Welchman for her fiercely relentless friendship and guidance. Assistants Jorge Vindas (Costa Rica), Marisa Montebella (Sardinia), and Kadowaki Kunio (Okinawa) were the unseen engines behind our successful stories.

No project of this magnitude happens without sponsors and financial partners. I wish to especially thank Marty Davis, the Davis family, and DAVISCO for their commitment to health and vast generosity; Jane Shure from the National Institute on Aging who was instrumental in obtaining our initial funding from the National Institutes of Health; Becky Malkerson, John Helgerson, Laura Juergens, and Maria Lindsley who championed Blue Zones at Allianz Life; Valerie May and Nancy Graham for navigating the waters at AARP; Nishino Hiroshi who found most of the funding in Japan; the Target Foundation, the Best Buy Foundation, Lawson Software, and the National Geographic Expeditions Council.

At Blue Zones' Minneapolis headquarters, Scott Meyer has been our mentor and marketing guru from the very beginning. The office team: Matt Osterman, Sarah Kast, Phil Noyed, Amy Tomczyk, Nancy Fuller McRae, and

Jennifer Havrish have endlessly helped with research, proof-reading, and have patiently endured my nonlinear methods; and the extended team including PR maven Laura Reynolds; Remar Sutton, Dr. Mary Abbott Waite, and the late George Plimpton, who provided crucial editorial assistance; Britt Robson for his help on the Okinawa and Loma Linda Chapters; our advisors including Tom Rothstein, Frank Roffers, Elwin Loomis, Jon Norberg, Ed McCall, Tom Gegax, Kevin Moore, Molly Goodyear, Chris Mahai, John Foley, and John Gabos who lent generous business advice; Thad Dahlberg, Dan Grigsby, and Bruno Bornstein, who built the Blue Zones website; and Keiko Takahashi, who created the Blue Zones identity.

And to the members of the media who have taken a chance and made the Blue Zones a national story I'd like to thank: Diane Sawyer, Rob Wallace, Jennifer Joseph, Anderson Cooper, Barbara Walters, Sanjay Gupta, Alyssa Caplan, Ned Potter, Patty Neger, and especially Walter Cronkite.

And finally, to writer Stephanie Pearson, who helped me flesh out this book during hundreds of phone conversations over the past seven years.

Preface

Get Ready to Change Your Life

THE FIRST TIME I MET SAYOKO OGATA, SHE was wearing the sort of fashionable gear one might expect a young Tokyo executive to take on a safari: hiking boots and cuffed socks, khaki shorts and shirt, and a pith helmet. Never mind that we were in Naha, a high-tech city of 313,000 on the main island in Okinawa, Japan. When I gently poked fun at her by saying that I could see she was ready for adventure, she didn't blush. Instead, she responded with one of her joyous, staccato laughs, wagged a finger at me, and scolded, "I'll get even with you, Mr. Dan." But I never saw the pith helmet again.

At the time, in the spring of 2000, Sayoko was a young, fast-climbing executive in Tokyo. Her company had brought me to Japan to explore the mystery of human longevity, a topic that would likely spark the imagination of a large audience. For more than a decade, I've been leading a series of interactive, educational projects called "Quests," in which a team of Internet-linked scientists investigated some of Earth's great puzzles. Our goal was to engage the imaginations and brainpower of tens of thousands of students who followed our daily dispatches on the web. Previous Quests had taken me to Mexico, Russia, and throughout Africa.

I'd first learned about Okinawa's role in longevity studies a few years earlier, when population studies indicated it was among the places on our planet where people lived the longest, healthiest lives. Somehow Okinawans managed to reach the age of 100 at a rate up to three times higher than Americans did, suffered a fifth the rate of heart disease, and lived about seven good years longer. What were their secrets to good health and long life?

I landed in Okinawa with a small film crew, a photographer, three writers, and a satellite technician to keep us connected to about a quarter million school kids. We identified gerontologists, demographers, herbalists, shamans, and priestesses to contact, as well as centenarians themselves, who were living emblems of Okinawan longevity.

Each morning our online audience voted to decide whom we should interview and where the team should focus its research. Each night we reported back to the audience with a variety of dispatches and short videos.

Sayoko had brought a team of translators with her, a computer filled with spreadsheets, and an intimidating plan to make sure that our daily reports and videos were translated into Japanese and transmitted by midnight to

Tokyo. We spent ten hectic days asking questions about life on Okinawa and summing up what we found. I met lots of fascinating people, which made me happy. Sayoko made her deadlines, which made her happy. And when the project was over, her team and mine celebrated with karaoke singing and sake and then parted ways. That was that.

THE BLUE ZONES QUEST

Five years later, I returned to Okinawa with a new team of experts. I'd just written a cover story for NATIONAL GEO-GRAPHIC about the "Secrets of Long Life," which profiled three areas of the world with concentrations of some of the world's longest-lived people—areas we dubbed "Blue Zones." Demographers had coined the term while mapping one of these regions on the island of Sardinia. We expanded the term to include other longevity pockets around the world. Okinawa still ranked among those at the top of the list.

I was determined to delve deeper into the lifestyle of Okinawans as part of a new online expedition—the Blue Zones Quest. More than a million people a day would follow our progress online. It was a huge opportunity to make a difference, and I knew we couldn't miss any deadlines. I decided to track down Sayoko.

She wasn't easy to find. I tried her old e-mail address and queried all of my old teammates concerning her whereabouts. I contacted her former boss, who told me she'd left her high-powered job behind to become a full-time mom. This news blew me away. By now I expected her to be a senior executive at Sony or Hitachi. Instead, she'd left Tokyo, he said, and moved to the island of Yaku Shima, where she lived with her husband, a schoolteacher, and their two children. When I phoned her, she was ebullient.

"Mr. Dan!" she said. "It makes me happy, really, to hear your voice." I told her about my new project in Okinawa and said I hoped she could join us.

"Dan," she replied, "you know I loved Quest, and for me it was really something quite powerful in my life. But now I have two small children, and I cannot leave them."

We talked for a few more minutes and then I hung up, disappointed. I'd have to find someone else. But a few days later, she called me back and abruptly accepted the offer. I had no idea why. I was just relieved to have her back on the team.

We set up our Blue Zones headquarters in a traditional guesthouse on the remote northern half of Okinawa. I'd recruited a team of scientists, writers, video producers, and photographers, and Sayoko had arrived with a team of Japanese translators and technicians. Gone was Sayoko's fashionable expedition wear. Now she wore sandals and earth-toned cottons. A few strands of gray streaked her hair, and she exuded calm. But when she opened her computer to a spreadsheet, I could see she'd lost none of her organizational zeal.

"Okay, Mr. Dan, let's talk about our deadlines."

For the next two weeks, we rarely saw each other face-to-face. During the day, my team gathered information and produced stories. Each night Sayoko's team translated them and published them to the Web. Since I was waking up about the time she was going to bed, we saw each other only at dinnertime, when both of our teams—20 of us—ate together. Our midnight deadline dominated all the dinner conversation. Sayoko and I never really got around to catching up on our personal lives.

LIFE CHANGES

Midway through the project, our online audience voted for my team to travel to Ogimi, a tiny fishing village, to interview a 104-year-old woman named Ushi Okushima. Sayoko and I had visited with her before when I had profiled her in my NATIONAL GEOGRAPHIC article. She'd impressed us with her amazing vigor, saying she grew most of her food and hosted drinking parties for her friends. Since turning 100, she'd somehow become a media darling. It seemed like every major news organization in the world, including CNN, the Discovery Channel, and the BBC had come to see her.

When Sayoko heard about our plans to visit Ushi again, she asked to come along. On the hour-long car ride to Ogimi, we had our first opportunity to really talk. We were sitting in the back seat as the vivid foliage of northern Okinawa zipped by.

"You know Dan, Ushi really changed my life," she began. "I'd been working in the center of Tokyo. I'd go from 7:30 in the morning until late at night, five to seven meetings a day, then dinner and karaoke until one or two in the morning. It was hard work, and I loved it. I did a good job. I made lots of money. But my life lacked something. I felt empty right here." She brought her hand to her chest.

"Dan, you remember," she recalled, "when we met Ushi I first saw her big smile. You were a man from another country, but she talked to you like a friend. In Japan, we're usually wary of strangers. Ushi immediately welcomed you. The atmosphere was like a big hug. You could tell that she made everyone happy around her—her family, her friends, and now even strangers. And even though she never even talked to me, I felt a big energy from her."

After our first meeting with Ushi, Sayoko said, she had gone into the street to drink some juice. "I was thinking, 'This is something for me.' For the rest of our trip in Okinawa, I thought of Ushi—the simplicity of her life, how she made people around her feel good, how she was not worried about getting something in the future or sad that she had missed something in the past. Gradually I was starting to think, 'I want to be like her. That is my goal.'

"When I returned to Tokyo, I told my boss that I was quitting. My dream had always been related to business. But I realized that I was like a horse chasing a carrot. I realized that I wanted to be like Ushi. I thought, 'How can I organize this?' I called my boyfriend in Yaku Shima and told him I wanted to visit. I moved to Yaku Shima and learned to cook. A year later, we were married.

"When my first child was on the way, my husband and I came back to Okinawa to meet Ushi again. I wanted her to bless my child. I don't think she remembered me. But my baby was born healthy. Now I have two children, and they are my life. No one knows about my career in Tokyo."

By this point, we'd almost reached Ogimi on a road that ran parallel to the sea. "What have you done to be like Ushi?" I asked.

"I've learned to make my own meals for the family," Sayoko said. "I put love into my food. I care for my husband

and my children, the husband comes home, and I have a good family. Also, I try to mentally check to make sure that I haven't hurt anyone, that the people around me are okay. I take time each night to think about the people around me, and think about what I eat, and what is important to me. I also do this during dinner. I take time to reflect. I'm not chasing the carrot any more."

RETURN TO USHI

By the time we arrived at Ushi's house, it was mid-afternoon. She lived in a traditional Okinawan wooden house with a few rooms separated by sliding rice paper doors and tatami mats on the floor. We removed our shoes and stepped inside. Though it is customary to sit on the floor, Ushi sat queen-like and serene on a chair in the middle of the room. When I first met her, she'd been anonymous. Now she had become a celebrity—a sort of "Dalai Lama" of longevity. Wrapped in a blue kimono, she motioned for us to sit down. So like kindergartners around a teacher, we sat cross-legged at her feet. I noticed that Sayoko, however, barely entered the room. For some reason, she seemed reluctant to get too close to Ushi.

By way of greeting, Ushi raised her arms above her head as if to show off her biceps and shouted, "Genki, genki, genki," or "Vigor, vigor, vigor!"

"What a treasure," I thought. So many people fear getting older. But if they could see this vibrant woman, they'd look forward to it. I showed Ushi the photograph of her in NATIONAL GEOGRAPHIC. I was beaming with pride that the story I'd written had made the cover. She looked at it blankly, put it down, and offered me a piece of candy.

I interviewed Ushi again, asking about her garden, her friends, and how things had changed in the five years since we'd last visited with her. She'd cut back on gardening some, she said, but had taken a job bagging fruit at a nearby market. She still spent much of her day with her grandchildren and the three surviving women in her circle of friends she'd had since childhood. She still ate a dinner of mostly vegetables and drank a cup of mugwort sake before bed. That was her secret, she told me. "Work hard, drink mugwort sake before bed, and get a good night's sleep."

As I spoke with Ushi, I caught Sayoko's eye. She was sitting off to the side, watching my interview. "Sayoko," I said, conscious of the fact that I was raising my voice inappropriately, but also figuring that Sayoko was too polite to approach Ushi without being beckoned. "Don't you want to tell Ushi your story?"

Sayoko hesitated but then came forward and knelt in front of Ushi. "Five years ago I came here, and you changed my life," she said. "Because of you, I decided to quit my job and get married. I owe you a big debt of thanks." Sayoko's eyes welled up with tears as she spoke. Ushi looked bewildered. She didn't remember their meeting.

"Then I came back a few years later," Sayoko continued. "You touched my belly when I was pregnant." This recollection now sparked recognition. Ushi smiled and then grabbed Sayoko's hands. Her thumb caressed Sayoko's thumb. "You inspired me, and now I am very happy," the younger woman said. "I had to come to thank you." Speechless, but understanding, Ushi patted Sayoko's hand. "I share my blessings with you," she said.

On the street outside Ushi's house, I caught up with Sayoko, who looked dazed but serene. I asked her what she was thinking. She smiled. "I feel like something is a little bit closed," she said in her own poetic Japanese-tinged English. "I feel complete."

CENTENARIAN WISDOM

This book is about listening to people like Ushi who live in the world's Blue Zones. The world's healthiest, longest-lived people have many things to teach us about living longer, richer lives. If wisdom is the sum of knowledge plus experience, then these individuals possess more wisdom than anyone else.

So we've packed this book with insights from centenarians about living life well. Their stories cover everything from child rearing to learning how to be likable, from getting rich to finding—and keeping—love in your life. From them, we can all learn how to create our own personal Blue Zones and start on the path to living longer, better lives.

When it comes to the science of living longer, however, centenarians can no more tell us how they reached age 100 than a seven-foot man can tell us how he got to be so tall. They don't know. Does Ushi's nightly cup of sake infused with mugwort provide some healthful benefits? Perhaps, but it doesn't begin to explain why she doesn't have cancer or heart problems or why she possesses such vigor at age 104. The way to learn longevity secrets from people like

Ushi is to find a place where there are many Ushis—to find a culture, a Blue Zone, where the proportion of healthy 90 or 100-year-olds to the overall population is unusually high. Then science can kick in.

Scientific studies suggest that only about 25 percent of how long we live is dictated by genes, according to famous studies of Danish twins. The other 75 percent is determined by our lifestyles and the everyday choices we make. It follows that if we optimize our lifestyles, we can maximize our life expectancies within our biological limits.

When we first set out to investigate the mysteries of human longevity, we teamed up with demographers and scientists at the National Institute on Aging to identify pockets around the world where people live the longest, healthiest lives. These are the places where people reach age 100 at rates significantly higher, and on average, live longer, healthier lives than Americans do. They also suffer a fraction of the rate of killer diseases that Americans do. We worked with some of the world's top longevity experts to distill lifestyles into the characteristics that could help explain their extraordinary longevity.

LONGEVITY LESSONS
This book begins by tackling the realities of aging. What are the chances that you will actually reach 100? What promises do supplements, hormone therapies, or genetic intervention offer? What are some of the scientifically proven ways for you to increase your years of healthy life?

In the following chapters, we'll take you to the world's confirmed longevity hotspots, the Blue Zones themselves:

the Barbagia region of Sardinia in Italy, Okinawa in Japan, the community of Loma Linda in California, and the Nicoya Peninsula in Costa Rica. In each of these places we'll encounter a different culture that has taken its own unique path to longevity. We'll meet longevity all-stars like Ushi and the experts who study their lifestyles and cultures. We'll show how history, genes, and time-honored traditions conspire to favor each population. We'll tease out the lifestyle components and let science explain why they seem be adding good years to people's lives.

The final chapter boils it all down into nine lessons, a cross-cultural distillation of the world's best practices in longevity. This, we believe, amounts to a de facto formula for longevity—the best, most credible information available for adding years to your life and life to your years.

Of course this information will do you no good unless you put it into practice. So, leading behavior experts will also offer an action plan to put these longevity secrets to work in your own life. And here's the good news: You don't have to do it all. We present an à la carte menu of sorts. You can pick and choose the most appealing items, follow our advice for converting items from the longevity menu into everyday habits, and know that whatever you choose, chances are you'll be adding months or years to your life.

Encoded in the world's Blue Zones are centuries—even millennia—of human experiences. I believe that it's no coincidence that the way these people eat, interact with each other, shed stress, heal themselves, avoid disease, and view their world yields them more good years of life. Their cultures have evolved this wisdom over time. Just as nature selects for characteristics that favor the survival of a species, I believe that these cultures have passed on positive habits over time in a way that most favors the longevity of their members. To learn from them, we need only be open and ready to listen.

Sayoko was ready to listen. Her brief time with Ushi led to a transition, helping her change from being a chronically stressed, marginally healthy professional to becoming a more serene, physically fit person living a life that matches her values.

Maybe you're ready to listen too. Who knows? It may change your life just as profoundly.

1

The Truth About Living Longer

The Truth About Living Longer

You May Be Missing Out on Ten Good Years

W HEN JUAN PONCE DE LEÓN LANDED ON the northeast coast of Florida on April 2, 1513, he was searching, it's been said, for a Fountain of Youth—a fabled spring of water that could bestow everlasting life. Historians now know there was more to the story. The reason the Spanish explorer set out to investigate lands north of the Bahamas was probably because Spain had reinstated Christopher Columbus's son Diego as military governor, effectively removing Ponce de León from the office. Nevertheless, the legend behind Ponce de León's voyage stuck.

The idea of discovering a magic source of long life still has so much appeal today, five centuries later, that charlatans and fools perpetuate the same boneheaded quest, whether it comes disguised as a pill, diet, or medical procedure. In an all-out effort to squash the charlatans forever, demographer S. Jay Olshansky of the University of Illinois at Chicago and more than 50 of the world's top longevity experts issued a position statement in 2002 that was as blunt as they could fashion it.

"Our language on this matter must be unambiguous," they wrote. "There are no lifestyle changes, surgical procedures, vitamins, antioxidants, hormones, or techniques of genetic engineering available today that have been demonstrated to influence the processes of aging."

The brutal reality about aging is that it has only an accelerator pedal. We have yet to discover whether a brake exists for people. The name of the game is to keep from pushing the accelerator pedal so hard that we speed up the aging process. The average American, however, by living a fast and furious lifestyle, pushes that accelerator too hard and too much.

This book is about discovering the world's best practices in health and longevity and putting them to work in our lives. Most of us have more control over how long we live than we think. In fact, experts say that if we adopted the right lifestyle, we could add at least ten good years and suffer a fraction of the diseases that kill us prematurely. This could mean an extra *quality* decade of life!

To identify the secrets of longevity, our team of demographers, medical scientists, and journalists went straight to the best sources. We traveled to the Blue Zones—four of the healthiest corners of the globe—where a remarkably high rate of the longest-living people manage to avoid many of the diseases that kill Americans. These are the places where people enjoy up to a 3 times better chance of reaching 100 than we do.

In each of the Blue Zones, we used a survey developed in collaboration with the National Institute on Aging to identify the lifestyle components that help explain the area's longevity—what the inhabitants choose to eat, how much physical activity they get, how they socialize, what traditional medicines they use, and so forth. We looked for the common denominators—the practices found in all four populations—and came up with what I consider to be a cross-cultural distillation of the best practices of health, a de facto formula for longevity.

Herein lies the premise of *The Blue Zones*: If you can optimize your lifestyle, you may gain back an extra decade of good life you'd otherwise miss. What's the best way to optimize your lifestyle? Emulate the practices we found in each one of the Blue Zones.

LONGEVITY PIONEER

In 1550, Italian Luigi Cornaro wrote one of the first longevity "best sellers." His book, *The Art of Living Long*, said that life could be extend through practicing moderation. His book would be translated into French, English, Dutch, and German. Cornaro may have been on to something; sources differ on his exact age, he lived well into his 90s and possibly beyond.

FACTS ABOUT AGING

When taken together, the Blue Zones yielded nine powerful lessons to achieve a longer, better life. But before we get into the details, I think it's crucial to understand a few things about just how people age and establish some basic principles and definitions. How long can each of us expect to live? What really happens to our bodies when we age? Why can't we just take a pill to extend our lives? How can we live longer? How can we live better? And why does changing our lifestyles add more good years?

To answer these and other fundamental questions, I've asked some of the world's experts to describe their latest research in everyday terms. Together these scientists represent the best thinking in biology, geriatrics, and the science of longevity.

Steven N. Austad, Ph.D., studies the cellular and molecular mechanisms of aging at the University of Texas Health Center at San Antonio. A professor at the Sam and Ann Barshop Center for Longevity and Aging, he is the author of *Why We Age: What Science is Discovering About the Body's Journey Through Life.*

Robert N. Butler, M.D., is President and CEO of the International Longevity Center-U.S.A., a policy and education research center in New York City. A professor of geriatrics and adult development at Mount Sinai Medical Center, he is the author of *Why Survive: Being Old in America.*

Jack M. Guralnik, M.D., Ph.D., is chief of the laboratory of epidemiology, demography, and biometery at the National Institute on Aging in Bethesda, Maryland.

Robert Kane, M.D., is director of the Center on Aging and the Minnesota Geriatric Education Center at the University of Minnesota in Minneapolis. He is a professor in the

School of Public Health, where he holds an endowed chair in Long-term Care and Aging.

Thomas T. Perls, M.D., M.P.H., is director of the New England Centenarian Study, an associate professor of medicine and geriatrics at the Boston University School of Medicine, and author of *Living to 100: Lessons in Living to Your Maximum Potential at Any Age.*

I interviewed each of these experts separately, then sorted the best of their answers to each question. Here's what they told me.

WHAT EXACTLY IS AGING?

Robert Kane: That is a very profound question. Number one, aging starts at birth. If you think about it, there is a constant development that occurs within all species. You can think of it as the balance between the individual and the environment. In essence, we can think of aging as a loss of coping mechanism, a failure to be able to maintain internal control and balance.

We start out as children, and we gradually accrue various changes in our characteristics. Children are susceptible to the environment and must be protected. In the case of humans, we probably peak in our mid-20s. We hold our own for a while, then at some point, perhaps in our mid-40s, we start to decline. Some people would say we actually begin to decline at age 30. It depends on the system that you track.

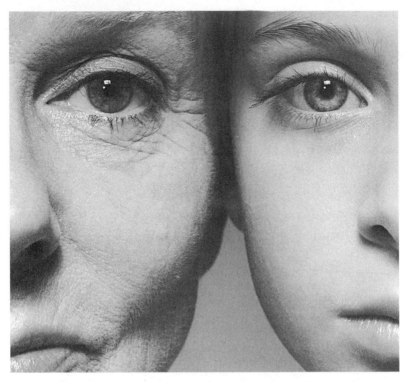

The skin is often where we first notice the outward signs of the aging process.

Old age is another period when the balance favors the environment; older people need help in protecting themselves. The frailty we associate with old age is basically the loss of autonomy, the inability to withstand external pressures and perturbations.

Aging includes both the positive and negative changes that occur. A gerontologist would define aging as the risk of dying. Irrespective of the presence of disease, there is, given this finiteness of a life span, a continuous risk of dying. In most cases this increases as our age increases. Other factors can change your risk of dying as well as aging, so it's not that aging alone is the determiner, but it is the overarching change. People have been searching for biological markers of aging, and so far nobody has found any that are absolutely constant and separate from the onset of diseases.

People look at, for example, the loss of accommodation in the lens of the eye. Most people become farsighted, usually in their early 40s. It doesn't happen to everyone, so you can't say it's a universal sign of aging. Graying of hair, loss of collagen in the skin, all of these are changes that have been described with regard to aging. There's a change in body composition as people get older. It can obviously be influenced by exercise and diet, but in general, we lose muscle mass and gain body fat. The immune system changes with age and becomes less competent, but again, not in any universal way that we can say is a characteristic of aging.

Steven Austad: I would define aging as the gradual loss of physical capabilities, whether you're talking about the ability to run, to think, all those things. It's a gradual and progressive loss of physical and mental abilities, the ability to do things that you previously could do. What it means is that basically, you're not designed to maintain your physical integrity forever.

Robert Kane: There are several theories about aging. One is that there are genes in your system that turn on and turn off, either to ameliorate or to expedite aging. Another theory is the "Garbage-Dump Syndrome," which theorizes that you accumulate toxins as you go along and things happen.

But again, the question you have to ask is why does the body accumulate toxins? Well, you probably accumulate toxins because some of the intracellular mechanisms that were working at one point have stopped working. So are the toxins really a sign of aging or merely a concomitant of some other biological process that has changed, presumably driven by some genetic clock that exists inside the body? We honestly just don't know.

WHAT IS THE AVERAGE AMERICAN LIFESPAN?

Robert Kane: I would imagine that a 30-year-old person today has a reasonable chance of living—depending on whether a man or a woman—to his or her late 70s or early 80s. If you took away major risk factors such as heart disease, cancer, and stroke, you would be adding, I would guess, maybe 5 to 10 years to that initial life expectancy.

Tom Perls: For most of us, our bodies are like cars built to go 100,000 miles. A few cars can go 150,000 or more miles with the right genetic makeup. But they do deteriorate over time, even with the best upkeep. With that deterioration comes frailty. When you hit a bump in the road, you are less capable of bouncing back. There comes a point, with

PEARLY WHITES

A vital part of your digestive system, a bright smile can last a lifetime. Healthy teeth allow us to eat the wide variety of foods found in a balanced diet, but cavities, ill-fitting dentures, and other oral problems can make chewing painful and may sometimes even lead to eating disorders. Be sure to make regular dental visits and brushing and flossing a priority to keep your smile healthy.

continued decline, where there's no bounce back, and that's when you pass away.

WHAT ARE THE CHANCES OF LIVING TO 100?

Jack Guralnik: Well, they are small of course, probably less than one percent. Again, figuring it out depends on what age you currently are. If you're talking about someone at birth, it's a different estimation than for someone who's already made it to 80. Also, if you want to consider health status, that plays a large role. Most people who make it to be centenarians when you look back, they were quite healthy at 80.

Tom Perls: I used to equate living to 100 to picking all five numbers in the lottery: The odds are pretty small. If you have longevity running in your family paired with good health behaviors, your chances are greater.

Centenarians today are the fastest growing segment of our population, partly because we're doing a better job of screening for high blood pressure. That's one important lottery ball that we don't need to leave to chance. Now instead of five numbers, it's down to four.

Another one that we've pretty much gotten rid of is substantial childhood mortality. With much better public health measures like cleaner water supplies, more years of education, improved social-economic status, these things are all reducing the number of lottery balls.

The best way to think about reaching 100 is: "The older you get, the healthier you've been."

Ponce de León's legendary search for the Fountain of Youth is one of the many tales that illustrate the desire to overcome aging.

Steven Austad: The question is—and here's where I think the best health practices are really important—if you live to be 100 years old, what sort of 100-year-old are you going to be? Are you going to be bedridden and unable to take care of yourself? Or are you going to be reasonably independent and alert? To me, that's what the best health practices can really have an impact on.

IS THERE A PILL THAT CAN EXTEND LIFE?

Robert Kane: There are a lot of nostrums out there. None of them has credibility. None of them has been even close to rigorously tested, everything from to human growth hormone to antioxidants. Every time anyone has studied them with any degree of rigor, they do not pan out. That does not mean that some new discovery may not be just over the horizon, but at the moment that is probably not the path.

Just think about it: If antioxidants were so healthful, the whole generation that grew up eating Twinkies, Wonder Bread, and the like (the kinds of foods that are loaded with antioxidants to assure that they had a long shelf life and would never spoil) should never grow old.

LEADING KILLERS

•Heart Disease: For both U.S. men and women, the leading cause of death
•Cancer: The second largest killer of American men and women
•Prevention: The Centers for Disease Control and Prevention advocates practicing a healthy lifestyle and regular medical check-ups and screenings

Robert Butler: DHEA, human growth hormone, and melatonin are all extremely questionable, and are probably ill-advised. Using human growth hormone in human beings bulks them up. But it does not just mean more muscle mass. With it can come hypertrophy of the heart, fluid retention, and other problems. And of course there's a disease, acromegaly, which is actually characterized by an excess of human growth hormone. DHEA is what's for years been called the "junk hormone." In large quantities in our bodies, it converts to both testosterone and estrogen. Most of the studies on almost all of these hormones have been very short-lived, that is six months to a year. So the long-term effects are not well known.

The best source of information on hormones is Marc Blackman at the Washington, D.C., Veterans Affairs Medical Center, or Mitchell Harman at the Kronos Longevity Research Institute in Phoenix, Arizona. They've done the most sophisticated studies and probably the best ones we presently have on the hormone story.

ARE VITAMIN SUPPLEMENTS HELPFUL?

Robert Butler: Of course you should maintain your basic daily vitamin requirements. But you shouldn't get carried away either. Vitamin E was under study by the National Institute on Aging in the hopes that it would prove to be very valuable with Alzheimer's disease. But it was not.

So I think, like so many things in nature, it's a matter of amount, what might be called proportionality, or just plain wisdom. People used to think if a multivitamin was good for them, then more of it would be even better, but that's just not true, unfortunately.

Most vitamin requirements are best achieved by eating six to nine servings of fruits and vegetables a day. Very few people do that, so probably the cheapest, least expensive multivitamin you can buy is not a bad idea to help achieve them. If you're an older man, you should not have a supplement with iron because iron accumulates in the heart and can lead to a condition called hemosiderosis. Look on the market for vitamin supplements that do not have iron that are designed specifically for men.

WHAT'S A SMART DIET FOR LONGEVITY?

Robert Kane: Eating a reasonable diet makes a lot of sense. Again, it doesn't mean that I think you have to be a vegetarian. One of the goals to a healthy lifestyle is moderation in all things. What one is looking for is moderation, taking in a level of calories that is necessary and balancing those calories across carbohydrates, fats, and proteins. Taking in really what you need. There are some things we know that are just generally bad. Most fast foods are not necessarily healthy. We seem to like a lot of the things that are bad for us: salt, sugar, fat. There is something about humans that is inherently self-destructive, at least when it comes to eating.

The best diet is basically one of moderation. You hear about all these people that live on legumes and plant foods and that's probably okay, but I don't think it's necessary. One certainly can metabolize a certain amount of meat, but again it's a question of are you eating European portions or American portions? Are you eating meat a couple of times a week, or are you eating it every day for two meals a day? Are you eating processed meats that are filled with fat? Or are you eating good cuts of fairly lean meat?

To me, I just come back to moderation. Assuming that you were in pretty good shape in your 20s, if you could maintain that weight, you would be in good shape. The truth is at 20, you could for all sorts of reasons, eat all sorts of terrible things and maintain that weight, because you were more physically active, because your system was just more resilient. As you get older, you lose that resilience. So you are more susceptible to lifestyle behaviors that can do you harm than you were when you were younger.

WHAT CAN ADD ON MORE GOOD YEARS?

Robert Kane: Rather than exercising for the sake of exercising, try to make changes to your lifestyle. Ride a bicycle instead of driving. Walk to the store instead of driving. Use the stairs instead of the elevator. Build that into your lifestyle. The chances are that you will sustain that behavior for a much longer time.

And the name of the game here is sustaining. These things that we try—usually after some cataclysmic event has occurred, and we now want to ward off what seems to be the more perceptible threat of dying—don't hold up over the long haul. We find all sorts of reasons not to do it.

The second thing I'd tell you is don't take up smoking. The biggest threat to improving our lifestyles has been cigarette smoking. That trumps everything else. Once you're a nonsmoker, I would try to get you to learn to develop a moderate lifestyle in regard to your weight to build into your daily routine enough exercise to keep you going.

DOES GOING TO THE GYM HELP?

Robert Kane: Exercise has several quite distinct functions. You have cardiovascular exercise, which we describe as aerobic exercise, which increases your body's capacity to process oxygen. That's where you go out and work really hard and raise your heart rate. Swimming would be a good way to get that kind of exercise.

There's also antigravity exercise. For example, if you're trying to prevent osteoporosis, swimming isn't the optimal activity, because it doesn't increase the strength of your bones.

Frank Shearer first put on water skis in 1939. At age 99, he still enjoys the activity.
"I like the outdoors and the exercise," he says.

There, working against gravity, walking, standing does more to increase bone metabolism than swimming does.

Then there are exercises designed to improve your balance. Tai Chi is one people talk about, or yoga. Those are exercises that have been associated with reducing the risk of falls.

Then there are strength exercises, which run all the way from weightlifters, who probably put themselves into a disadvantageous state from overdeveloping their muscles, to people who do some modest amount of weightlifting or antigravity exercise that strengthens their muscles.

The data suggest that a moderate level of sustained exercise is quite helpful. There are studies that show that people who run marathons tend to have much better cardiovascular systems than people who don't. You could say that that says more is better, but those exercises generally take a toll on your joints. So marathoners have good cardiovascular systems, but they will probably have to have their joints replaced. But in general, if somebody could do a minimum of 30 minutes—maybe we could raise it to 60—of exercise at least five times a week that would help. And it doesn't appear to have to be all at one time, although that seems to be better. If you did that and you could sustain it, that would be good.

HOW CAN WE MAXIMIZE THE GOOD YEARS?

Robert Kane: Again, there are two issues here. How long can I live? The other is: How well can I live? And those are different questions. Living an extra two years on life support may not necessarily be your goal. The question is: Can you delay the onset of disability? "Good years" is a very important concept.

There are some things I'd certainly recommend for what people would call successful aging. One of them is, in fact, to have a sense of social connectedness. Most people enjoy the company of other people, particularly other people who they feel care about them. That seems to give you a sense of well being, whether that raises your endorphin level or lowers your cortisol level. We don't know why. People have looked for biological markers, and they haven't been successful at finding them. But something happens that makes life more worthwhile. The days take on more meaning.

The other thing that helps a lot of people is doing something they feel is either interesting or worthwhile. Again, different people have different things they like to do. For instance, people talk about workaholics as being at higher risk for stress-related illness. But there is no evidence that workaholics are necessarily a higher risk if they really are enjoying what they're doing. If they are driven by some externality and feel like they have to earn more money, it creates stress in their lives, which is probably not very healthy. So it's very individual when it comes to what people want to do.

SMOKING & SKIN

The most preventable causes of death and disease in the U.S.A. are caused by smoking tobacco. In addition to the damage done to internal organs, smoking also prematurely ages the skin and makes people look older. Recent studies have shown that a smoker's skin bears more wrinkles and other signs of premature aging. The causes are still under investigation.

For example, you can't just say family support is good, because some family support is good for some people, and some isn't for others. There are people who derive great satisfaction from being with their families. And then there are those who become very anxious and upset when they are with their families. It is a complex model, which is also very interactive.

But if we're talking about things that give you a sense of fulfillment, a good life, the sense of being valued, the sense of being cared for, and the sense that you are liked—these are all very positive.

Tom Perls: A good start to adding more good years to your life would be to get rid of the anti-aging quackery.

Some people provide this very pernicious, ugly view of old people that's completely false in order to get you worried about getting older. They say they can stop—and even reverse—aging, claims which are absolutely false. You've got a bunch of people who are professing to be physicians or scientists, who are saying that they can stop or reverse the aging process. I will tell you that real scientists cannot do such a thing, so what makes the public think that these people can?

It is mostly hucksterism and charlatanism. They will cost you a lot of money, and these things do not work and, in some instances, can be bad for you. So stay away from it. These guys are just trying to sell you stuff. What does work is living the lifestyle of those who we know are living longer, like those people, I suppose, living in the Blue Zones.

INTO THE BLUE ZONES

Which brings us back to the Blue Zones project. Over the course of seven years, my team circled the globe, making several trips to each of the four Blue Zones and meeting the remarkable people who lived there. In each place we confirmed that people were as old as they said they were, interviewed dozens of centenarians, worked with local medical experts, and methodically studied each of the local lifestyles, habits, and practices.

Each Blue Zone revealed its own recipe for longevity, but, as we were to discover, many of the fundamental ingredients were the same. These common ingredients, our nine lessons of living longer, are deeply embedded in the cultures we studied. I suppose you could say that our quest was for a true fountain of youth, though this fountain does not spring from the ground but comes to us through centuries of trial and error.

For us, it all began on a small island off the coast of Italy.

2

The Sardinian Blue Zone

The Sardinian Blue Zone

Where Women Are Strong, Family Comes First,
and Health Springs from the Rugged Hills

N OCTOBER 1999, A SLIGHT, BESPECTACLED
Italian doctor and medical statistician named Gianni
Pes stepped to the podium at an international longevity conference in Montpellier, France, and presented an
astonishing paper. For the previous five years, he reported,
he'd traced the history of 1,000 Sardinian centenarians,
personally examining about 200 of them. During the
course of his research, he'd noticed a curious concentration of male centenarians in the mountainous Barbagia
region, a kidney-shaped area in the administrative district
of Ogliastra. The population there appeared to be the
longest-lived in Italy, perhaps even in the world.

In one village of 2,500 people, he said, he'd found seven
centenarians—a staggering number, given that the ratio

for centenarians in the U.S. is roughly one per 5,000. "All of the demographers attending the meeting were skeptical," Pes recalls. They remembered all too well the longevity claims made decades ago about populations in Georgia in the Soviet Union, in Pakistan's Hunza Valley, and in Ecuador's Vilcabamba Valley, which had all turned out to be overstated and based on faulty data. "I had a hard time convincing them," he recalled.

RECORD SETTER

Born on February 21, 1875, Jeanne Calment lived for a record-setting 122 years, 164 days. Calment, a French-woman, stayed mentally and physically active for most of her life. She attributed her longevity to port wine, olive oil, and a sense of humor.

Among those in attendance was Dr. Michel Poulain, a Belgian demographer who'd dedicated much of the past 15 years to studying pockets of long-lived peoples around the world. Among other things, he'd helped devise a widely accepted technique for verifying ages, which he'd used successfully throughout Europe and parts of Asia.

"I did not believe it," Poulain told me later. "The number of centenarians in Sardinia was too high to be true. I suspected misreporting. But someone nominated me to go check it out. As it was, I was traveling to Italy anyway, and so I said OK, I will have a look."

A REPUTATION FOR LONGEVITY

By the time he arrived in Sardinia in January 2000, word had reached the village of Arzana about its growing reputation as a place with an unusually high number of centenarians.

Residents in the village had organized a ceremony honoring four of them. "They invited me to participate, but I had no data," Poulain said. "I could not publicly confirm that theirs was a long-lived village. I am a scientist, and with no data, no conclusion.

But a half hour before the ceremony, I stopped by the city hall and looked at the birth and death records, and right away I found some preliminary indications that these documents were very accurate. I began to believe Gianni's findings, and so I participated in the ceremony and decided to go on with a large study to prove the exceptional male longevity in Sardinia."

Three months later, Poulain returned to Sardinia for the first of ten visits to check more records and interview centenarians. He visited 40 municipalities to establish something called the Extreme Longevity Index (ELI). This index considered birth and death records of all centenarians born between 1880 and 1900. Little by little, he realized that this region had a phenomenally high index. As he zeroed in on municipalities that had the greatest numbers of long-lived people, he circled the area on a map with blue ink—giving rise to the term "Blue Zone," which was later adopted by demographers.

Four years later, Poulain, Pes, and six colleagues published a paper in the journal *Experimental Gerontology*, in which they identified the Barbagia region as one that had some of the longest-lived people in Sardinia. The Blue Zone phenomenon primarily affected men, they reported. These men appeared to retain their vigor and vitality longer than men almost anywhere else. In most developed parts of the world, women centenarians outnumbered men four to one. Here, the ratio was one to one.

Their study had shown that the geographic distribution of longevity in Sardinia was not homogeneous. In at least one geographic area, the Barbagia Blue Zone, the probability of becoming a centenarian was higher than in other parts of the island. This area of extreme longevity was located in a mountainous region that had been relatively isolated until recent times. "The specific mechanism by which persons living in this territory were more likely to reach extreme longevity remains unknown," the researchers concluded.

As a demographer, Poulain's work was describing populations with data, not jumping to conclusions. Other than pointing out that other areas of extreme longevity existed in mountainous areas, his paper did not attempt to explain why the inhabitants of this Blue Zone were able to live so long. Unraveling this mystery would require a multidisciplinary approach to the history, diet, and lifestyle of local populations, he believed. Did people in this region experience stress? And if so, how did they shed it? Did religion play a role? Traditional medicines? Pure air? Something in the water? Did the Blue Zone hold any secrets that might help the rest of us live longer?

BLUE ZONE'S BEGINNING

In October 2004, National Geographic photographer David McLain and I landed in Sassari, a university town near Sardinia's northwestern coast, to look more deeply into this

Blue Zone's mystery. Two young Italian journalists, Gianluca Colla and Marisa Montebella, went with us to help set up interviews, translate, and handle logistics. Our plan was to interview at least 20 centenarians who personified the Blue Zone culture of longevity. From these interviews we would distill cultural characteristics, then meet with local experts who could help explain why these characteristics might contribute to the extraordinary longevity on the island.

Luckily for us, a Stanford University–trained evolutionary anthropologist named Dr. Paolo Francalacci— "Please, call me Paolo"—was teaching genetics at Sassari University. Cutting the image of a dashing young professor in his sporting blue jeans, tweed jacket, shirt open wide at the collar, and longish brown hair the day I met him, he led me through the narrow cobblestone streets of the town of Alghero, past a large piazza with a fountain and outdoor cafés, to a dimly lit 400-year old bar. We sat on benches at a corner table and ordered a couple of beers. A man of incandescent intelligence, Francalacci was one of those rare academics able to explain complex concepts in simple, compelling terms, often delivered with an auctioneer's exuberance and wild gesticulations.

He had first become interested in human evolution as a biology student at the University of Pisa, he said. That interest led him to join the laboratory of renowned geneticist Luigi Luca Cavalli-Sforza at Stanford, where he studied human populations by looking at their genes. His specialty was analyzing mitochondrial DNA to identify the origins of peoples—dead or alive. He had examined mummies found in China's Taklimakan Desert and revealed that they were of Indo-European origins, a discovery that had brought him fame.

"We have 46 chromosomes, half from our mother and half from our father," he explained, hands flying about like those of an orchestra conductor. "That means that for each gene we receive two copies, one from each parent, and these two copies interact. This is not true for two small pieces of our DNA: the Y chromosome, inherited from male to male (the females do not have it), and the mitochondrial DNA, inherited from female to female (the males do have it, but they cannot pass it to their offspring). This peculiarity makes it much easier to trace back the history of a population through the female (in the case of mitochondrial DNA) or male (Y chromosome) to its founding ancestors. Using DNA, we've traced every human being on Earth back to founding female lineages."

Francalacci had used this technique on Sardinians to trace their origins back roughly 14,000 years, he said. At that time, the world was warming after an ice age. As the snows retreated, a small band of genetically related people in Iberia began a journey out of the Pyrenees Mountains to the Mediterranean Sea. They followed the coast eastward, through what is today the French Riviera and Tuscany, and across the sea to Corsica, where they stayed briefly. Finally they settled in Sardinia's coastal foothills.

"Eighteen thousand years ago, during the period called the Glacial Maximum, humans could survive in Europe only in two refuges, one in Iberia and one in the Balkans," he said. With the retreating of the glaciers and warming of the climate, people started to repopulate Europe. They

Photographed almost a century ago, this child and three Sardinian women dressed in traditional garb gather outside a doorway in 1913.

moved westward from the Balkans and eastward from the Iberian Peninsula. Sardinia was populated almost exclusively by the Iberian wave—people with the M26 lineage of Y chromosome.

"This M26 genetic marker is found in 35 percent of the Sardinians today, and is very rare elsewhere," Francalacci said. Given the genetic purity of people in this Blue Zone, he theorized that the first Sardinians did not intermarry much with other Mediterranean peoples. They probably survived by hunting, gathering, and fishing. Agriculture came to Sardinia about 6,000 to 7,000 years ago with a Neolithic people from the Levant, where agriculture had been developed at least 3,000 years earlier.

"Our Y-chromosome data suggests that contact with these people from the Levant was mainly cultural rather than genetic," Francalacci said. For that reason, the people of Sardinia remain genetically distinct from the rest of Europe. Some of their unique traits are negative, such as higher incidences of type 1 diabetes and multiple sclerosis. But others are positive, such as resistance to malaria and high longevity rates, especially among males.

When we finished our beers, I needed a break. Interviewing Francalacci was a bit like witnessing a verbal volcano; I hadn't had to prompt him at all. He invited me back to the small rooftop apartment he shared with his lovely Greek wife, Christina. The view from their kitchen window looked out across a jumbled plateau of red clay shingles to the sea.

Francalacci popped a disk of Barbagia folk music in his player and the apartment filled with the nomadic, faraway sounds of Sardinia's highlands—shepherds' voices harmonizing to the multi-pipe instrument called *launeddas*. The musician plays it by producing a constant airflow by inhaling through his nose and exhaling out his mouth.

Francalacci opened a bottle of Sardinian Cannonau red wine, and his animated conversation resumed.

During the middle Bronze Age, a tribal culture called the Nuraghic civilization (in a sense, the root culture of the Sardinian Blue Zone) began in Sardinia, he said. The Nuraghi people are named after the stone towers found all over the island.

By the time of Christ, other civilizations had also discovered Sardinia's riches and charms, and for most of its early history the island was knocked around like a rugby ball—invaded, conquered, and exploited by outsiders. The Phoenicians and Romans arrived with their superior military might, taking over the coasts and lowlands of the south. Native Sardinians, who had lived throughout the island, escaped to the hilly central area. By most accounts, the invading barbarians were nomadic and largely interested in tending their flocks.

The etymology and meaning of the word Barbagia derives from the Latin name "Barbaria," land of the Barbarians. Latins called a foreigner *barbarus,* from the ancient Greek word *barbaros,* which supposedly mimics the sound of someone trying to speak Greek. They had no interest in the arduous tasks associated with agriculture, although they possibly learned rudiments of farming from the Romans. "Even if ancient Sardinians knew of farming techniques, it didn't take," Francalacci said. "They carried on largely as hunter-gatherers and later as shepherds."

Perhaps that's why Sardinians developed an intense wariness and disdain for visitors. Newcomers had always meant subjugation, exploitation, and taxes. So they turned inward, developing an intense dedication to their families and community and earning a reputation for toughness. One Barbagia proverb said it all about foreigners: *Furat chie benit dae su mare* (He who comes from the sea is here to steal).

As the centuries, passed on the island, the isolated native Sardinians evolved in their own unique environment. Many villages even retained their pre-Roman names. "In this region," says Francalacci, "the names of the Sardinian villages such as Ollolai, Illorai, Irgoli, Ittiri, Orune, to name a few, sound very exotic to a continental Italian ear like mine." The region north of Alghero is called Nurra, which some linguists think comes from *nur,* meaning "heap of stones" in the Nuraghian language. It is also a good description of the *nuraghi,* the Bronze Age towers found all over Sardinia.

The original Sardinians, in fact, did not keep their ancient Nuraghic languages. The Romans had subjugated them long enough that by the time they escaped to the mountains they had adopted Latin, which has survived the centuries remarkably intact. In the Sardinian dialect spoken in the Blue Zone, for example, the word for house is still the Latin word *domus.* Their pronunciation more closely resembles Latin too. The English word sky is *cielo* in Italian but it is *kelu* in Sardinian, preserving the hard K sound as it was pronounced in the original Latin *caelum* (ka-AY-lum). The same goes for sentence structure.

A modern-day Italian says *io bevo vino* (I drink wine) but Sardinians would say it as an ancient Roman would have, *io vino bevo* (I wine drink).

What does this have to do with longevity? "It suggests that the Sardinians' lifestyle in the Blue Zone hasn't changed much since the time of Christ," Francalacci said. "The laws of evolution dictate that a species will not evolve in a comfortable, isolated environment where reproduction is easy. By contrast, a species will evolve quickly in a tough environment where individuals of different backgrounds and conditions interact. Similarly, in a place like the Sardinian Blue Zone, there is less pressure to adapt. The people there maintained not only their genetic features, but also their economic isolation and traditional social values, such as the respect for elders as a source of experience, the importance of the family clan, and the presence of unwritten laws—all of which proved to be effective means for avoiding foreign domination over the centuries."

In other words, the self-imposed isolation of Sardinia created a genetic incubator of sorts, amplifying certain traits and subduing others. Some preliminary genetic studies, for example, seem to show that the red blood cells of an unusually high proportion of Blue Zone centenarians are smaller than normal, providing both a resistance to malaria and a lesser chance of dangerous blood clots. But genetic and cultural isolation go hand in hand, Francalacci said, putting down his wine glass and clasping his hands. "When you put the two together you get a very interesting result."

Francalacci and I stayed up well past midnight that night and continued our conversation later via e-mail. Before

1950, I learned, Sardinia had looked more like a backwater than a centenarian's Shangri-la. Poor hygiene, poor water quality, and a scarcity of water led to rampant infectious diseases. Dysentery, plague, tuberculosis, malaria, and diarrhea killed many young Sardinians. As English traveler William Henry Smyth wrote in his 1828 *Sketch of the Present State of the Island of Sardinia,* "it is surprising that with such inconvenient residences, and uncleanly habits, the natives should remain so generally healthy as they do."

IS IT IN THE GENES?

When D.H. Lawrence traveled across Sardinia in 1921 in search of a lifestyle of simplicity, he found a Barbagia suited to his imagination. "Here, since endless centuries man has tamed the impossible mountainside into terraces, he has quarried the rock, he has fed his sheep among the thin woods, he has cut his boughs and burnt his charcoal, he has been half domesticated even among the wild fastnesses. This is what is so attractive. . . . Life is so primitive, so pagan, so strangely heathen and half-savage."

Prosperity came to Sardinia after the late 1940s. The Rockefeller Foundation financed an effort that wiped out malaria, and a post-war economic boom in Italy brought jobs and paved roads to Barbagia. Along with them came vaccinations, antibiotics, and modernized health care. Only now could the Sardinian Blue Zone's combination of genes and lifestyle work its real magic on the population.

Before we left Sassari, photographer David McLain and I picked up a few more useful bits of information. Dr. Luca

Deiana, a local politician who headed the Akea study, one of the first investigations of Sardinian centenarians (*akea* is a Sardinian greeting that means roughly "may you live to be 100") pointed us in the direction of the G6PD gene. A defect on the gene is connected to favism, a disease triggered by consuming fava beans. We know the G6PD gene also protects some Sardinians from malaria, said geneticist Dr. Antonio Cao, who was himself a poster boy of healthy aging at 78. Diet probably plays an important role, he added. "Barbagia is not like the rest of the Mediterranean. They don't eat a Mediterranean diet."

Dr. Gianni Pes, the scientist who first delivered the Blue Zone data to demographers, also told us that environment and lifestyle might be more important factors than genetics to explain the longevity of Sardinians. "Consider, for instance, the genes of inflammation. We expected to find something interesting in Sardinian DNA. We studied several tens of gene variants related to inflammation but we didn't find any evidence of their role in survival of Sardinians. The same for genes related to cancer, and those related to cardiovascular disease. I suspect that the characteristics of the environment, the lifestyle, and the food are by far more important for a healthy life."

Armed with these insights and a backpack full of academic papers, David, Gianluca, Marisa, and I finally left the coast for Sardinia's Blue Zone. As we drove into the island's central highlands, we entered quite another world. The road snaked upward through increasingly rocky terrain that showed little evidence of human impact. Indeed, Sardinia consisted almost entirely of mountainous terrain as

"My tongue still works perfectly," says Raffaella Monne, 107. "I can talk a lot."

it rose toward the massive Gennargentu range to the east. Except for patches of hardwood forests where blackthorn, yew, oak, and ash trees grew, and the occasional vineyard, we saw only rough pastureland.

I recalled the many warnings we'd seen or heard about Barbagia. "You might think twice about wandering around many of the desolate villages of the interior, especially in Nuoro province," wrote the author of one guidebook.

Our friend Franco Diaz from Cagliari, Sardinia's largest city, confirmed our initial impression of Barbagia as a difficult place where people eke out a living from a rugged land by raising sheep and goats. The residents there have a reputation for kidnapping, stealing, and settling scores at the end of very long knives, he said. "A vendetta can last generations. A son of one family might get shot today for something his father did decades ago." Franco's daughter agreed. "If a boy catches you looking at his girl, expect to be confronted," she warned. "And remember, everyone in Barbagia has a knife in his pocket."

DUELING CENTENARIANS

As we drove into the village of Arzana on a drizzly October day, we saw smoke curling out of chimneys into the lingering mist. Villages at the heart of Sardinia's Blue Zone—Fonni, Gavoi, Villagrande Strisaili, Talana, and Arzana—have evolved over the centuries from clusters of shepherd huts to modern communities of a few thousand residents. Most towns still possess a precarious charm, with ancient whitewashed houses cascading down cobblestone streets. The streets of Arzana seemed all but deserted.

I could see why residents here might be physically fit. A trip to a friend's house or the local market meant a workout more rigorous than a half hour on a StairMaster. But the ominous creep of modernization was easy to see. Cars and trucks were parked in front of most houses, satellite dishes faced out from rooftops, and pizza, hamburger, and ice cream shops dotted the main street. To unlock Sardinia's longevity secrets, we needed to focus on Barbagia's traditional lifestyle—the one that existed before prosperity arrived in the 1950s.

Our plan was to track down as many as a dozen Blue Zone centenarians. Everybody in town knew them, we discovered, and people treated them like celebrities. On tavern walls, instead of posters of bikinied women or fast cars, you'd see calendars featuring the "Centenarian of the Month."

EQUAL EXERCISE

Everyday hikes taken by Sardinian shepherds can burn up to 490 calories an hour; to get the equivalent, try 120 minutes of brisk walking (about 3.5 mph), 90 minutes of gardening, 2 hours of bowling, or 120 minutes of golfing (be sure to carry your bag).

All we needed to do was ask a villager where the centenarians lived and a helpful finger would point out the house. We'd knock on the door, introduce ourselves, and invariably be invited in. During the first week, I met 17 centenarians—8 men and 9 women. Most centenarians, we discovered, spent their time somewhere between their bed and their favorite sitting chair. Their days were punctuated by meals with their families and perhaps a stroll to meet friends. As a rule, they had worked hard their whole lives as farmers or shepherds. Their lives unfolded with daily and seasonal routines. They raised families who were now caring for them. Their lives were extraordinarily ordinary—with one exception.

In Silanus, a village of some 2,400 people on the slope of the Gennargentu Mountains whose origins date to Nuraghic prehistory, David and I met 102-year-old Giuseppe Mura. We stepped out of the hard, hot midday sun into the 19th-century whitewashed home Giuseppe shares with his 65-year-old daughter Maria and her family. Inside it was cool and pleasantly damp; it smelled vaguely of sausages and red wine. Giuseppe sat at the end of an ancient, wooden table, flanked by Maria and his son, Giovanni, who'd stopped by for a visit.

Both father and son wore shepherd's caps, wool suit coats, and black boots—the daily uniform of the Sardinian peasant. Muted afternoon sunlight sifted through finely embroidered, diaphanous curtains. Giuseppe's eyes caught mine and he nodded agreeably.

"These men are from America," Maria shouted in her father's ear. "They'd like to interview you for NATIONAL GEOGRAPHIC magazine."

"What?" He shouted back.

"They want to interview you for a magazine article. The NATIONAL GEOGRAPHIC!"

"Fine with me," Giuseppe snapped. "But if they want money, tell them they can go to hell."

I blanched, but Maria and Giovanni burst out laughing. They understood their father's biting sense of humor, which was, I'd soon learn, characteristically Sardinian. I had some questions that Drs. Paul Costa and Luigi Ferrucci from the National Institute on Aging had given me to pose to centenarians. They were nonleading questions, carefully crafted to tease out the lifestyle by eliciting a narrative. Instead of asking a man what he ate when he was a child, the question would inquire, "Can you think about things you do every day or have done most days of your life?"

I posed my questions to Maria, who translated them for her father. I learned that Giuseppe had worked steadily his whole life, first as a farmer, and then as shepherd. He'd usually put in a plodding, 16-hour day tilling the earth or following his sheep into pasture. On most days he came home for lunch, took a nap, and then spent an hour or two in the late afternoon with his friends in the village square. He'd return to the fields until dark. He never much busied himself with raising his eight children. He left that and all other affairs of the house to his wife.

Giuseppe's diet consisted largely of fava beans, pecorino cheese, bread, and meat as he could afford it, which was rarely in the early days. Maria estimated that her father drank a liter of Sardinian wine every day of his adult life, and more during festivals, when he tended to be the life of the party.

"Is there anything unusual about Giuseppe's upbringing?" I asked.

Giovanni paused and looked questioningly at Maria. "Yes, there is," Giovanni answered. "Giuseppe was raised by a single mother. His father got his mother pregnant and then went off to war. When he returned, he took up with another woman and soon got her pregnant too, leaving my father's mother to have her baby alone."

Giuseppe listened to the story staring down at his folded hands, head hanging down. I could tell that even though more than a century had passed, the story was a source of shame for the family.

"Well, it looks like Giuseppe turned out all right," I added hopefully.

"Yes," Giovanni replied. "But there's more. One Sunday morning, when Giuseppe's father was on his way to church with his new wife, Giuseppe's mother intercepted him and shot him dead, murdered him right on the church steps. The police put her in jail, but everyone in the village knew that her honor had been violated. The police let her out after only four months."

What happened to the other woman and her child then? I was guessing that life was pretty difficult for a single mother in the village in first few years of the 20th century.

"That's another story in our village's lore," Maria said, taking up the tale. "The other child, Giuseppe's half brother, was named Raimondo Arca. Giuseppe did not even know he had a half brother until one day when he was 17 years old and was playing a Sardinian game with other boys in the village square.

"It was a game of elimination," she continued, "much like rock, paper, and scissors, but the Sardinian version often turns aggressive. Giuseppe and Raimondo found themselves in a final round of elimination when a dispute erupted and a fistfight ensued.

"Raimondo was giving my father a licking when a passing villager who knew the story of their mothers broke them up saying, 'Brothers shouldn't fight.' The secret was out, and once they learned it, they became friends and remained so ever since. In fact, Raimondo is still alive and lives right down the street. The whole village celebrated when then two of them turned 100 two years ago."

David and I looked at each other slack-jawed. In the United States only about one male in 20,000 reaches age 100. The chances that there would be two centenarians in the same family is astronomically unlikely, unless of course their father passed down an extraordinary set of genes to both of his sons.

The interview continued another 90 minutes. Maria served us wine and cured ham, followed by cups of hot coffee. In the course of our conversation, we learned that Giuseppe had a special box where he kept his meager life savings and other important objects that he had collected through his long life. He wore the key to the box on a string around his neck, which he closely guarded.

David, always on the lookout for a telling picture, envisioned a portrait of Giuseppe with his life's treasures. He asked Maria if it were possible for Giuseppe to open up the box for us.

"Papa," Maria again shouted in her father's ear. "These men want to see what's in your box."

"What?" Giuseppe murmured back.

"These men want to see what's in your treasure box, where you keep your money," she repeated, this time reaching for the key around his neck and holding it up for him to see.

"You tell those Americans to go to hell," he shrieked, slapping the key out of her hand. Spittle shot from his mouth and splattered on the table before me. "I show them what's my box, all right. Right up their nose!"

Even though Maria and Giovanni once again laughed at their father's outburst, we took this as our cue to leave.

WINE, GOAT'S MILK, AND MASTIC OIL

Though most centenarians we met were sharp enough to hold a conversation and answer questions, a majority of them were homebound and under the care of a daughter or granddaughter. What I learned directly from them was limited by their imperfect memories.

I realized that if I wanted to get a sense of the authentic Sardinian lifestyle, I needed to spend time with someone younger who was still working and living in the traditional way. I suspected that clues to the extraordinary longevity in this Blue Zone were the kind I needed to observe rather than hear described in an interview. I figured if I could participate in a day in the life of a true Barbagian Sardinian I could observe nuances.

FLAVONOIDS

Sardinian red wine isn't the only place to find flavonoids. Brightly colored fruits and vegetables and dark chocolate also contain them. Studies have shown that a diet high in flavonoids is associated with a reduced incidence of certain cancers and heart disease.

As it happened, photographer David McLain had already met such a person. While I had been working my way through interviews in the eastern part of the Blue Zone, David had been canvassing the western part for situations to illustrate our story (NATIONAL GEOGRAPHIC photographers and writers rarely travel together). He called me on the phone one afternoon to say he'd met a 75-year-old shepherd in the 3,000-year-old village of Silanus who still tended his own sheep, made his own wine, and lived in a traditional Sardinian home. His name was Tonino Tola, David said, "but I call him 'The Giant.'"

When I caught up with Tonino a week later, he was slaughtering a cow in the shed behind his house, his arms elbow-deep in the animal's carcass. A strapping, barrel-chested man, he withdrew his fist from the steaming mess and then strongly gripped my hand, vise-like, with a moist, bloodied handshake.

"Good morning," he boomed, then plunged his hands in again, this time to reel out several yards of glistening intestines. It was 9:45 a.m. on a cool November morning. Tonino had been up since 4 and had already pastured his sheep, cut wood, trimmed olive trees, fed his cows, and eviscerated this 18-month-old cow that was now hanging spread-eagle from the rafters. Members of his family surrounded him.

Tonino's son and three sons-in-law helped while his daughter cradled his five-month-old, wide-eyed grandson, Filippo, who regarded the scene with a cooing glee. With rolled-up sleeves and high-water pants, Tonino jumped from pulling cow's intestines to tickling his grandson with equal exuberance.

The men joked as they followed the age-old ritual of reducing the cow to meat for their family. The butchering was timed for late fall, when cooler temperatures minimize maggot-laying flies and make the meat easier to preserve. The cow would provide meat for two families for the season as well as for several friends who would receive gift packages of beef.

Butchering cows within city limits was illegal, Tonino told me, but this giant of a man lived by a more traditional Sardinian code. What would happen if the police caught

Shepherd Tonino Tola, 75, has climbed the hills of Sardinia for his entire life.
Regular physical activity is one reason why Sardinians live so long.

him, I asked. "We'll pay a fine," Tonino replied breath-
lessly, addressing a gaping cavity, the insides of which he
was scraping with a menacing blade, "or give him a piece
of meat."

Later I was invited into his low-ceilinged kitchen for
papassini—a Sardinian cookie made with raisins, almonds,
and a jam made from cooked wine (*saba*). Inside, a small fire
from a wood burning stove heated the room. Tonino's wife,
Giovanna, a heavyset woman with quick, intelligent eyes, sat

at the table. She offered us wine. ("No thank you" was not an option.) Tonino may have ruled the butcher shed, but Giovanna ruled the house. I posed questions to Tonino, and she answered them.

"Tonino lives to work," she told me, with her husky arms folded, "from early morning to late night. Look at him; he's aching to get back to butchering the cow right now." Sure enough, Tonino was drumming his fingers impatiently on the table; his eyebrows arched guiltily when his wife pointed at him. "Meanwhile, I take care of the house, the children, and the finances. I make sure we don't run out of money," she sighed. "He works, I worry."

I asked Tonino and Giovanna both about their past. They often finished each other's sentences. Tonino had raised sheep since he was five years old. He and Giovanna married when they were in their early 20s, and they quickly had four children. When their family was young, in the 1950s, they were very poor. They ate what they produced on their land—mostly bread, cheese, and vegetables (zucchini, tomatoes, potatoes, eggplant, and most significantly, fava beans). Meat was at best a weekly affair, boiled on Sunday with pasta and roasted during festivals.

They usually (Giovanna made the call) sold their animals to buy grain staples, from which they made their pastas and traditional breads—the flat, squarish *pistoccu* made with barley or bran flour and potatoes, and the famous lined, paper-thin *carta da musica* (also known as *pane carasau*) named for its resemblance to sheet music. Sheep and goat milk products contributed most of the protein. Their small vineyard grew Cannonau grapes for wine.

Their diet was fairly typical of families in the region before the American-style food culture arrived, as surveys from before the 1940s revealed. "Shepherds and peasants in Sardinia have an exceptionally simple diet, which is extraordinarily lean even by Mediterranean standards," a 1941 survey reported. "Bread is by far the main food. Peasants leave early in the morning to the fields with a kilogram of bread in their saddlebag. . . . At noon their meal consists only of bread, with some cheese among wealthier families, while the majority of the workers are satisfied with an onion, a little fennel, or a bunch of ravanelli. At dinner, the reunited family eats a single meal consisting of a vegetable soup (minestrone) to which the richest add some pasta. In most areas, families ate meat only once a week, on Sunday. In 26 of 71 municipalities surveyed, meat is a luxury eaten only during festivals, not more than twice a month. Interestingly for a Mediterranean culture, fish did not figure prominently into the diet."

The report went on to say that shepherds drank wine daily. "In the fields, some peasants drink wine; most of them drink wine only at the evening meal, and no more than a quarter bottle." The region's Cannonau grapes endured the harsh Sardinian sun by producing more red pigment to protect from the ultraviolet rays. These grapes traditionally were allowed to macerate longer than in any other part of the island during winemaking. The result was a red wine with two to three times the level of artery-scrubbing flavonoids than other wines.

Goat's milk and mastic oil may be Sardinia's other two longevity elixirs. Research at the University of Sassari is looking at the question of whether proteins and fatty acids in Sardinian goat milk may help protect people from the typical diseases of aging such as atherosclerosis and Alzheimer's disease. Mastic oil (with its antibacterial and anti-mutagenic properties) was used in some parts of Sardinia in place of olive oil.

Back at Tonino's house, we washed down a dozen cookies with a few glasses of wine. After a sedentary hour, Tonino could take it no more and erupted out of his seat. Almost every day for the past 70 years, he had walked or ridden his donkey the 5-mile journey to tend his sheep on his family's mountaintop pasture. But today, because he invited me, we would drive.

The road snaked up several hundred feet through forests and around tight curves and many unprotected drop-offs that promised a quick death. In America such a road would be illegal—or at least labeled "dangerous." Here it was business as usual.

We stopped at a high plateau fenced in by an ancient rock wall where 200 sheep had gnawed vegetation down to nubs. At the highest point of the pasture, a teepee-shaped rock-and-stick structure called a *pinnetta* commanded a 360-degree view of the property. In this structure, the likes of which dated back to the Bronze Age, Tonino slept with his sheep most summer nights. At the moment, though, he was looking jaunty in his leather spats, shepherd's cap, and riding coat as he strode through a narrow opening in a stonewall, counting his sheep as they followed him.

When three sheep tried to squeeze through, they knocked over a section of the wall. With disquieting ease, Tonino hoisted the heavy rocks back into place. Then he leaned back on a rock outcropping and assumed the age-old role of sentinel, cutting a distinguished profile against the emerald green plains below.

"Do you ever get bored?" I asked impulsively. Before the words left my mouth, I realized I'd uttered a heresy. Tonino swung around and pointed at me, dried blood still rimming his fingernail. "I've loved living here every day of my life," he boomed. After a pause, he continued. "I love my animals and taking care of them. We don't really need the cow that I butchered today. Half of the meat will go to my son, and most of the other half we'll share with our neighbors. But without the animals and the work it takes to raise them, I would be sitting in my house doing nothing; I would have little purpose in life. When I think of them, I think of my children. I like it when my kids come home and they find something here that I have produced."

THE POWER OF LOVING?

From everything I had seen in this Blue Zone, Tonino's values were also those of the region's population in general. People here possessed a reverence for family. Perhaps it had something to do with their historic isolation, surrounded as they were by hostile outsiders; they had to depend on one another. All the centenarians I met told me *la famiglia* was the most important thing in their lives—their purpose in life.

In America, seniors tend to live apart from their children and grandchildren, often sent off to retirement homes when they become unable to care for themselves. But that rarely happened here. A combination of family duty, community pressure, and genuine affection for elders kept centenarians with their families until death. This gave people over 80 a huge advantage: They received immediate care when injured or ill, and perhaps most significantly, felt loved and a sense of belonging. A happy by-product was that grandparents stayed involved in children's lives.

Maria Angelica Sale and her family were a perfect example of this. I met Maria, called Nona by her family, in Gavoi, one of the island's highest villages. She was sitting in her living room, a clean, well-lighted place with embroidered tablecloths, colorful carpets, and a large window that looked out onto the street. Her 60-year-old daughter, Pietrina, served me coffee and translated my questions from Italian into Sardinian, the only language most of the centenarians speak.

Maria had raised her four daughters, survived life under the Fascists, and worked alongside her husband from 4 a.m. to dusk. When her husband died, she moved in with her daughter at age 54. Then she helped raise her grandchildren, cooking food, cleaning house, and, until just two years ago, knitting their clothes.

On the day of my visit, Maria seemed subdued. Mostly blind, deaf, and unable to walk by herself, she was confined to a chair. But she was lucid, serene, and exuded a certain feeling of satisfaction with her life. Her gray hair was pulled back in a bun, revealing the soft wrinkles in her face. Leaning forward slightly to hear my questions, she would then sit back and seem to savor the memories they sparked before answering with a smile. The whole time, Maria's daughter held her mother's hands, which lay folded in her lap.

I asked Pietrina how her mother had managed to live so long, and she gave me a one-word answer: grandchildren. "It's about loving and being loved," she said. "Not only has Nona helped raise the children, but she has also always told them they must get educated. Sometimes she gives them money, and she always prays for them. In return the children have felt this love and have returned it. They know that Nona expects them to succeed, so they try harder."

Two years ago, when she was at age 100, Nona got very sick. "She was in bed for days," said Pietrina. "I thought she was going to die, so I called the family. Everyone came—4 daughters and 13 grandchildren—many of whom traveled back from the mainland. On the day we thought she was going to die, everyone had gathered around the bed to say goodbye. We didn't actually think she could hear us. But when my nephew, who was a failing student, leaned over to say how much he was going to miss her, Nona opened her eyes and said, 'I'm not going anywhere until you're done with the university.' Nona got better and my nephew went back and graduated."

It would be hard to overestimate the importance of family in the Blue Zone. According to Dr. Luca Deiana, who has studied centenarians for more than a decade, some 95 percent of those who live to 100 in Barbagia do so because they have a daughter or granddaughter to care for them. Grandparents provide love, childcare, financial help, wisdom, expectations, and motivation to perpetuate traditions and push children to succeed. This may add up to healthier, better adjusted, and longer-lived children, and it seems to certainly give the population a healthy bump in longevity.

"SARDONIC" IS RIGHT

But what is it about men in the Sardinian Blue Zone? Statistically, their story is the most extraordinary. In the Blue Zone, 47 men and 44 women lived past their 100th birthdays in a population of 17,865 born between 1880 and 1900—a rate of centenarians that exceeds America's by about a factor of 30. If the ratio of men to women among Sardinian centenarians was typical, the region should have produced only 10 male centenarians during the period. But it didn't. So why are men so extraordinarily long-lived here and not women?

Perhaps it has something to do with the different burdens they carry. The men were "quiet, and kind, and sensitive to the natural flow of life, and quite without airs," D. H. Lawrence observed, while women had "dangerous and hard assurance as they strode along so blaring. I would not like to tackle one of them."

Sardinian men seem to possess a temperament that enables them to shed stress. They are at once grumpy and likable, and often joke at the expense of one another. (It's probably no coincidence that the word sardonic has its roots on this island.) The things they have in common, according to physician Gianni Pes, are strong will, high self-esteem, and great stubbornness. "This is actually the national character of Sardinian men," he says, "and may well explain their success in surviving in unfavorable circumstances."

In Talana, one of five whitewashed villages that hang along Barbagia's coastal mountains like a string of pearls, Marisa and I met a 90-year-old shepherd named Sebastiano Murru. We had heard about his extraordinary vigor and had spent an afternoon searching for him. Early in the evening someone pointed him out to us standing at a bar with a half dozen friends, all much younger. He was wearing a shepherd's cap, tweed blazer, and black work boots. Though short in stature, he had an erect posture and his sly smile gave him a commanding presence. "Are you Sebastiano Murru?" Marisa asked, approaching him tentatively.

"I don't know. I was too young to remember when they named me," he replied matter-of-factly. Sebastiano's friends all laughed.

Then she asked how old he was.

"Sixteen," he responded, smirking. His friends erupted in laughter.

We noticed an empty beer glass in front of him, and asked him if he drinks. "No, my doctor told me not to drink. Especially not milk." He accepted a beer anyway and toasted my health. He was standing next to Marisa, who was 39 and attractive, so I returned the toast, recalling Groucho Marx: "May you always feel as young as the woman you're with." He looked at Marisa, examined her from head to toe and rejoined, "Do I look like a cradle robber?"

HUMBLED BY A HUNDRED-YEAR-OLD

At 103, Giovanni Sannai was good-natured, but lacked the edge. He lived alone in small house close to his son in the

Giovanni Sannai, 103, sits at the head of the table surrounded by his extended family.

village of Orosei. He received us at his kitchen table, where he greeted us with open arms and a tumbler of wine—which he insisted we drink with him. It was 9:30 in the morning.

With his husky frame and full head of graying hair, he projected the image of a man 30 years younger. He had an easy smile that made us feel sincerely welcome. I interviewed him for about two hours, during which time about a dozen neighbors stopped in to say hello. He greeted each of them with the same smile and tumbler of wine. Soon we had an audience.

He had an inexplicable quality that made you want to be around him; he was interesting and interested and up for new experiences. At one point, after he told me how strong he used to be, I gamely challenged him to an arm wrestle. He accepted. His friends surrounded us. I figured I'd just let him win. After all, I was 60 years younger.

I clasped his thick hand, fixed his gaze, and easily got the jump, taking his hand off-center. But it was a trick. He held my arm at about a 45-degree angle for several long moments until I tired and let up. Then he slammed my arm down on his side of the table. I couldn't have won if I had wanted to. Everyone laughed.

He confirmed many of the same things other centenarians had told me. He drank goat's milk for breakfast, walked at least six miles a day, and loved to work. For most of his life, he woke early and spent his day in the pastures. In the winter months, from November to April, he would herd his sheep over 100 miles to grassier lowland pastures. He made these journeys on foot, sleeping in *pinnettas* at night and eating only carta da musica bread, pecorino cheese, wine, sheep's milk, and the occasional roasted lamb—which they could obtain along the way. When I asked if he'd ever been stressed in his life, he looked flummoxed. I asked the question in a different way.

"Sometimes, but my wife was in charge of the house, and I was in charge of the field," he said. "What's there to worry about in the field?" Then he added, "Mostly I've always tried to remember that when you get good things from life, enjoy them, because they won't be there forever."

Gianni's words were poignantly prescient. Sardinians today have already taken on many of the trappings of modern life. Mechanization and technology have replaced long hours and hard work; cars and trucks have eliminated much of the need to walk long distances; a culture disseminated by television is replacing the one that put the emphasis on family and community; and junk foods are replacing the whole-grain breads and fresh vegetables traditionally consumed here. Young people are fatter, less inclined to follow tradition, and more outwardly focused (which also could lead to a dilution of this amazing gene pool). In 1960, almost no one in Sardinia's Blue Zone was overweight. Now 15 percent of adolescents are. The most important and unique longevity factors have disappeared or are disappearing quickly from residents' everyday lives.

SARDINIA'S LONGEVITY SECRETS

So what was the secret to longevity in Sardinia? Did it begin in the Bronze Age, when islanders retreated into the mountains, turned inward, intermarried, and developed a strong allegiance to community and family? Did that explain their traditional social values, such as the sense of hospitality, the importance of the family clan, the presence of unwritten laws, which were developed over the centuries? Perhaps their self-imposed isolation had created a genetic incubator of sorts that amplified certain traits and subdued others over the generations in a combination that favored longevity.

Gianni Pes thinks it might be something that Sardinians get, of all things, from goat's milk and cheese. He wrote to me recently to tell me that he's found a plant, Sardinian dwarf curry, that grows on the slopes of the Gennargentu Mountains and around the village of Arzana. An anti-inflammatory and bactericidal substance has been recently extracted from this plant.

GOAT'S MILK

When compared to cow's milk, goat's milk delivers a powerful nutritional punch: One glass contains 13 percent more calcium, 25 percent more vitamin B6, 47 percent more vitamin A, 134 percent more potassium, and 3 times more niacin. Results of a 2007 University of Granada study found that it may also be better at preventing iron deficiencies and mineral losses in bones.

"It represents one of the natural molecules of greatest interest as a new anti-tumor and anti-AIDS drug," Pes wrote. "It's also one of the most powerful anti-inflammatory drugs ever found. During my recent trip to Ogliastra, I spoke to a guy who told me that sheep do not eat this kind of plant, while goats do. Therefore one may ask whether goat's milk in this region is richer in arzanol, thus acquiring anti-inflammatory properties."

The region's unique geographic properties certainly provide a piece of the puzzle too. The Blue Zone's rocky, sun-beaten terrain, cut by deep valleys, was not suited for large-scale farming. So over the centuries, shepherding offered the best profession. The work was neither stressful nor strenuous, but it did require miles and miles of walking a day. (Walking five miles a day or more provides the type of low-intensity exercise that yields all the cardiovascular benefits you might expect, but it also has a positive effect on muscles and bones—without the joint-pounding damage caused by running marathons or triathlons.) Indeed, Sardinian male centenarians seemed to avoid bone loss and fractures. One Italian study has shown that Sardinian centenarians reported less than half as many fractures as the average Italian centenarian.

Meanwhile, Sardinian wives were more sedentary. Women here tended to stay home caring for children, doing home repairs, managing household finances, and worrying about their husband's safety. Unlike marriages in many Mediterranean cultures, Sardinian women wore the pants. They bore more than their share of the marital stress load, which, perhaps, enabled their husbands to live longer.

The Sardinian diet was lean and largely plant-based with an emphasis on beans, whole wheat, and garden vegetables, often washed down with flavonoid-rich Cannonau wine. Goat's milk and mastic oil, common in the diet 30 years ago, may also have provided powerful compounds.

Finally, for me, Sardinia's most important longevity secret lies in the unique outlook and perspective of its people. Their hardship-tempered sense of humor, which may seem caustic and persnickety to outsiders, helps them shed stress and diffuse feuds before they the start. Their fanatic zeal for their families has always protected them from a historically hostile world by providing cooperation in times of difficulty.

Consider the virtuous circle Tonino set up for himself: the relentless dedication to his family, the lifelong drive to work, and how those habits yielded protean health and robust vitality in his eighth decade. Remember also his capacity to take a few moments each day to admire the view of the island form his pastureland perch—though he's seen the same vista nearly every day for almost 80 years. How often do our hard-pressed lives allow us to take the time to appreciate the subtle beauty around us? Sardinians have the presence of mind to savor what they have—and perhaps they are calmed by this.

A final and subtle, yet powerful, Sardinian attribute is their positive attitude toward elders. In America, being young is celebrated, and growing old is often dreaded. I have heard that most seniors polled said they'd rather die than be put in an assisted living facility. But in America, we are increasingly having to care for our seniors outside the family: People over 65 in the U.S. will need long-term care for an average of three years, and more than a third of them will not rely on help from family.

There are no long-term care facilities in the Sardinian Blue Zone. There, respect increases with age. Younger generations feel an affectionate debt to the parents and grandparents who raised them. All but one of the 50 or so centenarians I interviewed had a daughter or granddaughter who actively cared for them. Remember Giuseppe? After the second time he suggested that David and I "go to hell," I gingerly commented to his daughter Maria that perhaps he'd be better off in a retirement home. "That would dishonor my family!" she retorted sternly. "Moreover, he would not be happy there."

Is there a connection between respecting elders and longevity? Absolutely. Seniors who live at home are more likely to get better care and remain engaged. In Sardinia, they are expected to help with childcare and contribute to the functioning of the household. They have strong self-esteem and a clear purpose. They love, and they are loved. And as we shall see in forthcoming chapters, purpose and love are essential ingredients in all Blue Zone recipes for longevity.

SARDINIA'S BLUE ZONE LESSONS

To live like a Sardinian centenarian, try the following practices.

Eat a lean, plant-based diet accented with meat.
The classic Sardinian diet consists of whole-grain bread, beans, garden vegetables, fruits, and, in some parts of the island, mastic oil. Sardinians also traditionally eat pecorino cheese made from grass-fed sheep, whose cheese is high in omega-3 fatty acids. Meat is largely reserved for Sundays and special occasions.

Put family first.
Sardinia's strong family values help assure that every member of the family is cared for. People who live in strong, healthy families suffer lower rates of depression, suicide, and stress.

Drink goat's milk.
A glass of goat's milk contains components that might help protect against inflammatory diseases of aging such as heart disease and Alzheimer's disease.

Celebrate elders.
Grandparents can provide love, childcare, financial help, wisdom, and expectations/motivation to perpetuate traditions and push children

to succeed in their lives. This may all add up to healthier, better adjusted, and longer-lived children. It may give the overall population a life expectancy bump.

Take a walk.
Walking five miles a day or more as Sardinian shepherds do provides all the cardiovascular benefits you might expect, and also has a positive effect on muscle and bone metabolism without the joint-pounding of running marathons or triathlons.

Drink a glass or two of red wine daily.
Tonino, Sebastiano, and Giovanni all drank wine moderately. Cannonau wine has two to three times the level of artery-scrubbing flavonoids as other wines. Moderate wine consumption may help explain the lower levels of stress among men.

Laugh with friends.
Men in this Blue Zone are famous for their sardonic sense of humor. They gather in the street each afternoon to laugh with and at each other. Laughter reduces stress, which can lower one's risk of cardiovascular disease.

3

The Blue Zone in Okinawa

The Blue Zone in Okinawa

Sunshine, Spirituality, and Sweet Potatoes

T HE SIMPLEST WAY TO IMAGINE OKINAWA is to envision it as a Japanese Hawaii—an exotic laid-back group of islands with a warm, temperate climate, palm trees, and sugar-sand beaches hemming a turquoise sea, where the cities ping and hum with an electronic din, and a favorite dish is a SPAM-and-vegetable stir-fry. For nearly a millennium, this Pacific archipelago nearly 1,000 miles from Tokyo has maintained a reputation for nurturing extreme longevity. Reports from Chinese expeditions referred to these tiny islands as the land of the immortals. Despite the ensuing years of Chinese, and then Japanese, domination, a devastating world war, famines, and typhoons, Okinawa can still claim to be home to some of the world's longest-lived people.

In February 2005, I made my second journey to the Okinawan Blue Zone to research a story for NATIONAL GEO-
GRAPHIC. I'd used Okinawa's incredible profile of longevity
to convince the editors that there were indeed pockets of
longevity around the world. In Okinawa people enjoy what
may be the highest life expectancy (in 2000 figures that
worked out to be 78 years for men and 86 years for women),
the most years of healthy life (the Japanese have the greatest number of disability-free years at 72.3 for men; 77.7 for
women), and one of the highest centenarian ratios (about
as high as 5 per 10,000). They suffer from diseases that kill
Americans, but at much lower rates: a fifth the rate of cardiovascular disease, a fourth the rate of breast and prostate
cancer, and a third the rate of dementia.

The economic impact of cardiovascular disease on the
U.S. health care system continues to grow as the population ages. The cost of heart disease and stroke in the United
States has been estimated at $432 billion for 2007, includ-
ing health care expenditures and lost productivity. Imagine
how much America might save if we could bring our heart
disease rates down to those of Okinawa.

Since lifestyle, not genes, is the chief determinant of how
long we live, I argued that the Okinawan Blue Zone offered
the world's best practices in health and longevity. During
my first trip in 2000, I spent time with 13 centenarians and
heard their stories. They seemed to eat lots of vegetables and
possessed a strong connection to their ancestors. Many were
prodigious gardeners, going into the fields every morning
and returning with tasty greens and tubers in the afternoon.
But did any of this explain their extraordinary longevity?

This time, I intended to take a closer look. I'd use a survey developed by the National Institute on Aging to systematically interview another few dozen Okinawans in search of common lifestyle characteristics. And I'd consult with scientists to find out how those characteristics connected to longevity. Dr. Greg Plotnikoff, one of those experts, agreed to travel with me. I'd met Greg years earlier at the University of Minnesota's Center for Spirituality and Healing, where he served as the medical director.

"In America we focus on battling diseases once they occur," says Greg, 46, who completed residencies in both internal medicine and pediatrics at the University of Minnesota, holds a divinity degree from Harvard, and is one of the world's leading experts on Kampo, Japan's traditional herbal medicine. "However, in traditional Asian thought, the highest, most honored form of medicine was prevention, and the lowest was treatment. Today in Japan, the focus is on avoiding disease in the first place. There are massive national and local efforts underway to prevent diabetes and heart disease. Japan's priorities represent a profoundly different way of understanding medicine."

At the University of Minnesota, Greg taught medical students the science of herbal medicines and dietary supplements as well as issues in cross-cultural clinical care. There, he was also the principal investigator of the largest clinical trial ever undertaken in the West of a Kampo formula. He obtained FDA approval to import and study in human volunteers *keishi bukuryo gan,* an 1,800-year-old remedy consisting of four herbs and a mushroom for treating menopausal hot flashes.

Recently he'd accepted an invitation to serve as an associate professor at Tokyo's Keio University School of Medicine, where he taught and studied the pharmaceutical efficacy of traditional medicines. His training, in my book, supremely qualified him to help us tease out Okinawa's lifestyle of longevity. No wonder I wanted him along.

AN UNLIKELY PARADISE

I joined Greg in Tokyo, spent a few days interviewing Japanese longevity experts, and then the two of us flew to Okinawa together. When we landed in Naha, Okinawa's capital city, we hailed a taxi and made our way through the slanting rain to the center of town. The streets were lined with a profusion of low-slung, typhoon-proof concrete structures that increased in height as we reached downtown. Towering electronic lights blinked on every building: Sony, Hitachi, Coca-Cola. As our taxi crept down gridlocked streets, raindrops splattered on the windshield in garish red, blue, and green blotches of refracted light. At an intersection near our hotel, we passed a bubble tea stand, a shop selling electronic merchandise, a Pizza Hut, and a McDonald's. I thought: This is the longevity Shangri-la?

The first morning, Greg and I caught up with photographer David McLain and assistant Rico Noce for breakfast in the hotel restaurant. David greeted me by pointing outside to rain that had fallen relentlessly through the night and snapped, "It's been like this for three days. I can't shoot in this." This was a bad omen. David, with whom I've worked on eight expeditions, has an irrepressibly good nature.

Okinawan Seiryu Toguchi, 105, soaks up a few moments of sunshine, which helps his body manufacture vitamin D.

I turned to Rico, who shook my hand energetically and demanded, "How are you?" We'd corresponded by e-mail, but I'd never actually met her in person. "Sit down," she instructed. Somewhat relieved, I could see she had the can-do pluck we needed to get the job done, rain or no rain. Over a breakfast of fermented soybeans, pickled cabbage, and raw fish, David, Greg, and I came up with a game plan. Meanwhile, Rico frantically took notes, occasionally stabbing at her breakfast of bacon and eggs.

Later that day, I'd arranged to meet Craig Willcox, a world renowned gerontologist and co-author of the *New York Times* best seller *The Okinawa Program*, at the hotel. In the early 1990s, Craig and his identical twin brother, Dr. Bradley Willcox, were doing a study on nutrition and healthy aging at the University of Toronto when they met a man who was to become the oldest in the province of Ontario. He was an immigrant from Okinawa named Toku Oyakawa, who at the time was 105 and still ate the diet of his homeland, fished every day, and kept his 92-year-old wife happy.

Toku got the Willcoxes thinking of two things: One, they wanted to see the real place that produced these long-lived people; and two, if an Okinawan living in freezing Canada could be healthy at the ripe old age of 105, what about the rest of us? Could Blue Zone behaviors travel?

In Okinawa, the Willcox brothers met Dr. Makoto Suzuki, a demure Japanese medical doctor who had discovered this Okinawan Blue Zone. He took the young scientists under his wing. In 1975, Dr. Suzuki had been sent to the main island from Tokyo to help open Okinawa's first medical school. While there, he'd heard about a famous centenarian living in the countryside and decided to investigate. Arriving at the centenarian's village, he realized that all he had was the woman's name, no address. "I saw this healthy-looking woman walking down the street with a hoe, looking like she was on her way to farm," Suzuki told me later. "So I stopped to ask her if she'd heard of the centenarian woman. She said, 'I am her.' I could not believe it! She looked like a woman of 70. But later I did the investigation and it was true."

The experience piqued Suzuki's curiosity enough to conduct a survey of the entire island. He identified 40 centenarians on Okinawa, an extraordinarily high number for such a small place. (Okinawa's main island is about 15 miles across at its widest point and a little more than 60 miles long.) But what really astounded him was the proportion of healthy centenarians. In the United States or Europe, you can expect only about 15 percent of centenarians to be independent in their activities of daily living. Although later cohorts of centenarians tend be less functional than earlier cohorts, amazingly, of the 32 centenarians Dr. Suzuki was able to track down for the second part of his survey, all but 4 were functionally independent.

During the past few years, the Willcox brothers and Suzuki had received ongoing international attention and they had grown weary of the endless interview requests and film crews. Luckily, I knew Craig from his precelebrity days, and he agreed to help me during my stay as much as his schedule permitted.

"Danmeister!" I heard him shout from across the hotel lobby. I was glad to see him, still the athletic academic, with a sleek physique, sharp features, and round, owlish glasses—a top-notch scientist with a boyish streak. I never knew how he got that name for me, but I let it stick.

"Craig," I shouted back. "You haven't changed."

I told him what I had discovered in Sardinia, and he told me about his findings over the past five years. He told me that the Okinawan

OKINAWA FACTS

• Nation: Japan
• Location: Ryukyu Islands between the North Pacific and the East China Sea
• Okinawa Population: 1.3 million

culture of longevity was beginning to disappear with the encroaching American food culture. Kentucky Fried Chicken and McDonald's were the last calamity to befall Okinawa; the fast-food invasion has threatened many of the positive behaviors that led to Okinawan longevity. It really rests with women over 70, he told me. "Men under 55 in Okinawa are now among the most obese, and do not live much longer than the Japanese average. So you better work fast." I asked him if he could help me meet a couple dozen centenarians as soon as possible.

"The privacy laws have tightened quite a bit since the last time you were here," he said. "It is going to be much harder for you to talk to centenarians and harder for me to help you."

"What about Dr. Suzuki?" I asked.

"Same thing," he replied.

The situation only seemed to worsen over the next two days. I went upstairs and brainstormed with Rico. If Craig Willcox, Dr. Suzuki, and the various government offices couldn't help make connections, the whole Okinawan part of the NATIONAL GEOGRAPHIC story might be a bust.

Rico took on the challenge. The government was reluctant to release names from the *koseki,* the family register system that is the Japanese equivalent of our birth registration system among other roles, but she had found a list of Okinawa's top 100 centenarians (there were some 700 among Okinawa's 1.3 million people) in the archives of a local newspaper. She figured we'd just look up their telephone numbers and start calling—but whether from hearing loss or tradition, many centenarians didn't have phones.

Those who did have phones frequently lived with protective children who were not inclined to grant access to foreigners. It wasn't a good way to communicate under the best of circumstances. We called one rather well-known centenarian, announced that we were from NATIONAL GEOGRAPHIC, and that we wanted to interview him. He replied that he was fed up with telemarketers and hung up on us.

A bright spot opened up on the third day when Craig phoned to say that he'd set up a dinner meeting for us that night with Suzuki at a nice sushi restaurant overlooking Naha's concrete expanse. It was loud, and the rice paper screen that separated our table from the main dining room did little to muffle the noise. Having just endured a fairly discouraging day, I sat down right beside Suzuki. I knew that he had personally visited more than 700 of the approximately 2,000 centenarians in Okinawa during his 25-year study; that he knew where many still lived, but would understandably be protective of their privacy. "Can you help us find centenarians?" I asked him directly.

"Very difficult," he replied in rudimentary English, although I know he can speak it fluently and understands even nuances of the language.

So he didn't know of any I could contact?

"No, no, can't do that!" he responded. "I can't give you any names, of course, but there is nothing that would prevent you from following me as I go visit these people . . . ."

I needed the hopeful boost Suzuki had provided. In Sardinia, I could go into almost any village in the morning and have three centenarian interviews completed by noon. Okinawa is different. Protocol required us to first receive the blessing of the mayor or another high-ranking official, wait a day or two for his office to arrange the interview (or not), and another day or two to facilitate our request.

I also knew that meeting centenarians was just part of the job. Piecing together the cultural recipe for how these centenarians lived *throughout* their lives was where the true answers lay. For that, I needed to get a feeling of how Okinawans lived before fast-food restaurants arrived on the island. Naha's concrete chaos didn't feel like part of that equation. But it was where I could find experts.

On day two, Rico and I visited Dr. Kazuhiko Taira, an M.D. and faculty member in the Department of Tourism Sciences at University of the Ryukyus. For 20 years he had studied residents of the Okinawan village of Ogimi and compared it to places in the Akita and Aomori prefectures in northern Japan. Every year he updated his survey of centenarians in Okinawa and found that Okinawans suffer significantly fewer strokes. He believed that it was related to diet, specifically the eating of less salt and more pork.

"Okinawa people are able to grow vegetables in gardens all year long," he said, referring to the island's comparatively tropical climate. "They don't need to make pickles and preserve food as people do in Japan's northern islands."

A salt-heavy diet may contribute to high blood pressure and weaken cerebral arteries, he said, causing micro-tears that are precursors to strokes. Consuming a balance of both animal and plant protein through pork may help repair or retard those tears. It has long been an Okinawan custom to butcher a family pig in celebration of the lunar New Year. The pig is cooked for a long time, the fat is skimmed off, and the rest of the animal is used in a stewlike dish.

"Okinawa people eat most every part of the pig," Taira said. "There is a lot of vitamin B_1 and B_2 in it, and collagen, which is good for you." This was different from many Japanese, who derive more of their protein from fish. Too much animal protein can increase your chances of obesity; the Okinawan centenarians traditionally ate meat only during infrequent ceremonial occasions.

THE POWER OF THE GARDEN

On our fifth day, I got an unexpected call from Craig. Dr. Suzuki and some medical students were planning to check on Kamada Nakazato, a 102-year-old woman living on the Motobu Peninsula. Would Greg and I like to join? You bet!

We drove in Craig's car, following Suzuki north past Naha's concrete jumble. We took the expressway along the island's backbone, past the enormous American military base and through Nago City. Then we turned onto

Motobu Peninsula, which on a map looks like a knobby appendage on the island's side. Here the roads narrow and the open spaces flanking them widen. We hugged the coast for a while and saw the blue, glimmering sea and a few beaches. In some areas, houses hemmed the coast, but semitropical vegetation fringed most of the shore. Inland, enormous vegetable gardens ringed the villages and old people wearing enormous conical hats bent over their crops. The air smelled like things ripening.

Kamada Nakazato's home was typical of those in the region—equal parts fortress and cozy enclave. Surrounded by a low, coral-rack wall, the house was somewhat sunken, with a steep, slanted roof—all architectural defenses against the half-dozen or so typhoons that blow in from the seas during Okinawa's typhoon season. Greg, who has crystal-blue eyes and an inclination toward bursts of enthusiasm, was craning his neck to take in the expansive gardens behind the house and motioning for me to do the same.

"There are herbs galore!" he said in a loud, faux whisper. "I'll bet her longevity secret is growing right out there."

We made our way inside the stone walls and onto the step-up deck in front. Rice paper screens served as a front door. Closed screens mean do not disturb; open screens are an invitation to the village to stop by for a visit. Kamada's screens were wide open. We called out a greeting and heard a yelp in reply. We took that as an invitation to enter.

Kamada was wrapped in a kimono and sat next to two of her children, both in their 70s. Her cottony, white hair was brushed straight back, revealing her high cheekbones and deep brown eyes, which widened as we walked through the door. When she saw us, she gleefully raised her hands and began singing, holding her hands up and swaying back and forth. Her children, taking the cue, clapped in unison. I felt an instant affection for her. She struck me as warm and approachable, yet possessing a commanding charisma.

Craig had told us in the car that Kamada was the village *noro,* a priestess who communes with the gods and ancestors and serves as spiritual adviser to the townspeople. Starting in the 15th century, the noro had become an official part of Okinawa's political structure, appointed by the court and assigned to certain villages. The succession of noro within a family became a basic rule, and the position was passed down to a niece, daughter, or granddaughter. Kamada was the last in her family's 400-year noro lineage.

Her regal title stood in sharp contrast to her home, a traditional, three-room structure made of weather-beaten wood. In one corner, an urn, some pine boughs, old family pictures, and a lunar calendar sat on what appeared to be an altar. Except for a bed, no furniture cluttered her house. Everyone, including Kamada, sat on the floor

Suzuki opened his medical kit and took her blood pressure, measured her percentage of body fat, and drew blood that he siphoned into little vials and put in his box for later analysis. He tested Kamada's mental acuity. "What year is this?" he asked.

"The year of the chicken," she shot back.

"What season is it?"

"That's a stupid question," she quipped. We all laughed.

"What day is it?"

Okinawan families honor their ancestors by having a meal at the family tomb.

"Today is the fifth of February, the thirteenth of the lunar calendar." Her noro duties required her to track festivals using the lunar calendar.

Satisfied, Suzuki invited me to ask a few questions. I asked her about her past, about what it was like being a young girl in Okinawa before World War II. Kamada's father, a rice and sugarcane farmer, could barely feed his family, so in the third grade, Kamada dropped out of school to help her mother raise the family.

"Life was hard," she began. "We had famines, times when people starved to death. Even when times were good, all we ate was *imo* (sweet potato) for breakfast, lunch, and dinner."

Her story matched what I had read of the island's history. Okinawan peasants like Kamada's family had lived hard lives. Most of them scraped out a living by cultivating millet, rice, and barley, which were poorly suited for the island's rocky soil. Though Okinawa's warm, semitropical climate provided for growing seasons, the half dozen annual typhoons that whipped through destroyed crops.

The Okinawan farmer and his family worked ceaselessly to keep up, and often suffered chronic malnutrition. The five-year-old child helped weed the rice field. The grandmother tended the vegetable garden. The grandfather carried wood from the hills. The idea of retirement never occurred to the Okinawan peasant. To this day there's not a word for it in their language.

Things improved a little in 1605 when an Okinawan brought the sweet potato back with him from China. This hardy miracle tuber thrived just fine in Okinawa's stingy soil and weathered its typhoons and monsoons. It was a boon for peasants, quickly becoming a staple. Boiled, it also provided food for livestock, so even the poorest Okinawan could now afford meat—albeit only during the annual lunar festival. By the time of Kamada's birth in 1902, Okinawans got 80 percent of their calories from sweet potatoes.

At age 18, Kamada entered into an arranged, political marriage with a man four years her senior. Their union produced three sons and three daughters. When the children were still small, Kamada's husband frequently traveled to mainland China and Palau in search of work.

"Raising children by myself was a big job," Kamada admitted. "At times there was no income. I had to weave straw hats to make money on the side." She pantomimed pushing a needle through thick straw to make a tight weave. I noticed that the repetition had permanently bent her right index finger at a 45-degree angle. Her husband eventually returned and they raised their children to adulthood. Two moved away but two still lived on the same street in rural Motobu. Her husband died ten years ago, at age 96.

The secret to surviving 75 years of marriage?

"I learned to be patient," she said.

I asked Kamada to describe her morning routine. "I wake up at about 6 a.m. and make a pot of jasmine tea and eat my breakfast—usually miso soup with vegetables. Then," she pointed toward her door, "I go to the sacred grove to pray for the health of the village and thank the gods for making it safe." I later learned that this sacred grove was a clearing in the forest about 650 feet (200 meters) from the house with a gazebo-like structure. Craig leaned over to me. "In her mind, she is very much convinced that the health of the village depends upon her paying attention to the stars and the moon and the spirits of her ancestors. Even as a centenarian, it is a job she undertakes very seriously."

At noon, Kamada said, she wanders into the kitchen garden behind her house to harvest some herbs and vegetables for her lunch. "I'll use mugwort to give my rice flavor or turmeric to spice my soup," she said. "I don't eat very much any more. Usually just stir-fried vegetables and maybe some tofu."

"And meat?" I asked.

SWEET POTATOES

Sweet potatoes are a delicious way to pack in vitamins and minerals. High in fiber, vitamin A, potassium, vitamin C, and folic acid, "sweets" are also easy to prepare. Prick one with a fork, microwave it for about five minutes, and just season with salt and pepper.

"Oh yes, I like meat, but not always. When I was a girl, I ate it only during the New Year festivals. I'm not in the habit of eating it every day."

"Have you ever eaten a hamburger or had a Coke?" I asked. I knew that Okinawa had more fast-food restaurants per capita than anywhere else in Japan; what may be the biggest A&W restaurant in the world was just 30 miles south of here. Perhaps she had visited one?

Kamada's forehead wrinkled. She leaned over to her daughter, for interpretation. "She never drank a Coke in her life," the daughter answered. When she first saw a hamburger a few years ago, Kamada had asked, "What do you do with that?"

"My mother eats in the tradition of women her age," the daughter continued. "They are not used to rich foods, but rather the foods that they ate as young women, before the war. She mostly eats vegetables from her garden—daikon, bitter melon, garlic, onion, peppers, tomatoes—and some fish and tofu. All day long she nurses a pot of hot, green tea. Before each meal she takes a moment to say *hara hachi bu*, and that keeps her from eating too much."

"Hara hachi bu?" I repeated.

"It's a Confucian-inspired adage," Craig chimed in. "All of the old folks say it before they eat. It means 'Eat until you are 80 percent full.' We write about it in *The Okinawa Program*." He continued, "Okinawa may be the only human population that purposefully restricts how many calories they eat, and they do it by reminding themselves to eat until they're 80 percent full. That's because it takes about 20 minutes for the stomach to tell the brain it is full. Undereating, as the theory goes, slows down the body's metabolism in a way such that it produces less damaging oxidants—agents that rust the body from within."

How about sweet potatoes? Did she eat them anymore? I was test-driving a hypothesis that sweet potatoes may be part of Okinawa's longevity theory. The tuber was so ubiquitous that before World War II, instead of saying hello, islanders greeted each other by saying, *Nmu kamatooin*, which translates as, "Are you getting enough imo?" They even erected a statue to the man, Sokan Noguni (now known as Lord Sokan or the Imo King), who first brought the sweet potato to the island. The tuber is incredibly healthy, extremely high in vitamin C, fiber, and beta-carotene—an agent shown to have cancer-fighting properties.

"No," she snorted. "I ate imo for breakfast, lunch, and dinner for 50 years. I got tired of it."

"No one really eats it anymore," Craig said. "It largely disappeared from the diet when the American food culture took over after WWII. It's hard to make a case for it being a longevity food."

Most afternoons Kamada naps, does some light gardening, and then, at 4 p.m. or so, joins a group of lifelong friends—her *moai*—for *sake* and gossip. She eats a very light dinner before 6 p.m. that might include some fish soup, whatever vegetables are in season, some spring onions, salad, and rice. She's usually in bed by 9 p.m.

"So what's the secret to living to age 102?" I ask, finally. I knew the question wasn't scientific, but sometimes it provoked insightful answers.

"I used to be very beautiful," Kamada replied. "I had hair that came down to my waist. It took me a long time to realize that beauty is within. It comes from not worrying so much about your own problems. Sometimes you can best take care of yourself by taking care of others."

"Anything else?"

"Eat your vegetables, have a positive outlook, be kind to people, and smile."

I looked over at Craig, who was sitting next to me, to see what he thought of her response. "Well, Danmeister," he said looking through his owlish glasses. "It took us almost 500 pages in our book to say what she said in three sentences."

Before I left, I wandered around Kamada's home. Her bedroom was furnished with only a thin futon rolled out on the tatami mats (thick, wall-to-wall pads made of rice grass). In her kitchen, a bamboo steamer hung from the ceiling just beneath an electric rice cooker. Small plates and tall narrow cups were neatly stacked and ordered on the shelves. There were no candy, cookies, or other tempting foods under the counter. If she had any junk food, it was hidden away in the cupboards.

Between Suzuki's tests and my interviews, we stayed with Kamada for about two hours, and then we let her rest. Ignoring a possible etiquette breach, I embraced Kamada before we left, which she enthusiastically returned. Beneath her kimono, I could feel her delicate bones. Her breath was warm on my neck. It occurred to me that I was embracing a century of life, and there was something in that knowledge that inspired profound affection and respect. I asked her if we could return. "Of course you may," she said. "I'm not going anywhere."

SOY AND A SENSE OF PURPOSE

After the interview, we stopped at a village restaurant for a late lunch. The meals were served in bento boxes, the Japanese equivalent of an individually sized picnic basket. I sat next to Suzuki and began plying him with questions, beginning with why he took Kamada's blood samples.

When compared to Americans of the same age, "we have found that the blood of male and female centenarians seems to have higher levels of sex hormones," Suzuki said.

"So is that why people are taking these hormonal supplements?" I asked, thinking about DHEA supplements, precursors to sex hormones (androgens and estrogens), which some believe can slow the effects of aging.

"Yes, but it doesn't work that way. The supplements are not the same as the hormones your body generates. DHEA probably has no effect on age-related changes in body composition and function."

Soy products that contain phytoestrogens are probably better than hormone supplements, he said. Some researchers speculate that they may impart many of the benefits of estrogen without the cancer dangers. Okinawans eat an average of three ounces of soy products per day. Tofu, their main source of soy, may play a role in reducing the risk of heart disease.

Greg Plotnikoff recommended that consumers select fermented soy products over nonfermented soy products whenever possible. "The medical literature demonstrates comparatively much better nutritional content in fermented soy," he said. "And Okinawan tofu has a greater concentration of protein and good fat than even its Japanese or Chinese counterparts."

I asked Suzuki if he thought Kamada's vitality came from hormones or something else.

"I think the fact that she still retains her duties as a noro is very important," he replied. "Roles are very important here in Okinawa. They call it *ikigai*—the reason for waking up in the morning. A sudden loss of a person's traditional role can have a measurable effect on mortality. We see this especially among teachers and policemen who die very soon after they quit working. Police and teachers have very clear senses of purpose and relatively high status. Once they retire, they lose both of those qualities and they tend to decline rapidly. I believe the reverse is true too. Kamada remains functioning longer because she feels needed."

CENTENARIAN CLUTCH

Rico and I returned to Motobu twice more to follow up on Kamada's story. I wanted to meet Kamada's moai and talk to the rest of her family. I was especially interested in how the younger generation regarded the rapidly disappearing older generation. Did they respect their elders? Was there any regard for the old ways in the age of fast food and text messaging?

I met Kamada's great-granddaughter, 14-year-old Kurara, in a village park where she was taking part in a school track meet. She blew away the competition in the first leg of an 800-meter relay race. As she passed the baton to her teammate, she threw her arms victoriously into the air and then ran to the finish line to cheer on her teammates. Kurara was still breathing hard when I caught up with her.

"Champions!" she shouted in her confident, deep voice, pointing to her teammates who smiled beside her. Kurara agreed to walk with me back to her great-grandmother's house and answer a few questions. I wondered what most impressed her about her grandmother.

"Straightforward." Kurara answered in a word. She was wearing a green T-shirt, white shorts, and Nike shoes; her short-cropped hair and healthy smile gave her tomboyish good looks. "Grandma doesn't keep stress. Sometimes she is so straightforward it could sound harsh. Like when we offer to take care of her, she'll say, 'No, I'll take care of myself!'" Then Kurara paused, realizing that perhaps she was sounding disrespectful. "I love my Grandma's sense of humor the most," Kurara added. "Sometimes she farts, and she tells me it was a train going by outside."

At Kamada's house, I took my shoes off outside the door and climbed gently onto the tatami mats. Kurara flew past me with her hand in the air, headed right to Kamada who, wrapped as usual in her regal kimono, sat serenely on a chair. "Give me five, Grandma!" Kamada raised her hand and slapped her great-granddaughter's. The she turned to me and beamed a smile.

Kurara told me about her family, her friends, and her affection for the Backstreet Boys. She said she loved watermelon, purple sweet potatoes, and *natto* (traditional fermented soybeans). She loved running and basketball and would rather be exercising than playing video games. When I asked Kurara what she wanted to be when she grows up, she puckered her lips and threw her head back. "I want to be a fashion model." She stood up and sashayed across the yard, swaying her hips as if she were on a runway. "I practice everyday."

"But I also love *kendo* (Japanese fencing)" she said. "I am a samurai." Switching roles in less than a second, she extended her imaginary sword as if she were ready to attack. "Those are my dreams. But I think I will probably become a physical education teacher. I want to teach kids to enjoy school. Mostly, though, I want to teach kids how valuable life is. I can introduce them to my grandmother so the students can hear stories of their lives. Of course I don't know if she will still be around."

I asked Kurara if she thought she would live to be as old as her grandmother. She looked at me with a straight face and said, "Of course. I am shooting for 150."

I waited a few days before returning to Motobu. It was late afternoon, and three days of rain hadn't let up. I had intentionally timed my visit for the gathering of her *moai*—the group of lifelong friends who meet at Kamada's every day. I walked past the coral stone wall and up to the house. Light shone through a rice paper screen, creating the silhouettes of several animated ladies. I stopped a minute to listen to the chatter of voices, followed by squealing laughter. I wanted to see the moai because I thought it might figure into Okinawan longevity.

SOY SHOPPING

Soy has the ability to lower the level of "bad" (LDL) cholesterol in the body and the potential to reduce the risk of heart disease. It can be found in a wide variety of forms, from tofu, to soymilk, to edamame (whole soybeans in their pods)—all reliable sources of soy's nutritional benefits. Protein content may vary among forms and producers, so be sure to check the labels.

The notion of moai—which roughly means "meeting for a common purpose"—originated as a means of a village's financial support system. If someone needed capital to buy a parcel of land or take care of an emergency, the only way was to pool money locally. Today the idea has expanded to become more of a social support network, a ritualized vehicle for companionship.

I meekly pulled aside the screen. Inside, Kamada's moai sat in a circle dimly lit in the glow of the kitchen light. Kamada—clearly the queen—sat in the center, coaxing heat from a small charcoal stove. She was now sitting in a large, wooden chair while the other four women sat primly on low stools or on the floor itself. "May I come in?" I interrupted. The group fell silent. Kamada raised her hands overhead to welcome me in.

For over an hour I sat in a corner and observed, with Rico whispering translations into my ear. The women gossiped and cracked jokes. Conversation ranged from romantic intrigue ("She stopped seeing him after she found him with someone else. Big surprise, eh?") to chatty news ("She got in a fight with her son-in-law because he is treating her daughter badly") to job postings ("My son needs some help with his market stall so if any of you have a hardworking grandson. . . .").

"Is this all about gossip?" I interrupted.

"No," replied 95-year-old Matsse Manna after a long pause. "If someone passes away, the village knows to come here for help. If we hear that someone is depressed we will go visit them."

"But how about you? How does this moai help you?"

"Chatting like this is my ikigai," said Klazuko Manna after a long pause. At 77, she was the youngest of the group. "In the morning I do the wash, so in the afternoon, I get to come here. Each member knows that her friends count on her as much as she counts on her friends. If you get sick or a spouse dies or if you run out of money, we know someone will step in and help." Klazuko fanned her arm toward the other women. "It's much easier to go through life knowing there is a safety net."

"I get lonely on days when the group doesn't meet," Kamada added. "I go to the door every day at 3:30, and if my friends don't come, I'm sad."

As I listen, it occurred to me that this little group of five women, who represented more than 450 years of life, might be the world's greatest repository of longevity wisdom. Japanese women on average were doing something special to live almost 8 percent longer than American women. Their moai may very well be part of the equation. Chronic stress takes its toll on overall health, and these women have a culturally ingrained mechanism that sheds it every afternoon at 3:30 p.m. Books like *Bowling Alone* chronicle how people in the United States are increasingly alienated from their neighbors. On average, an American has only two close friends he or she can count on, recently down from three, which may contribute to an increasing sense of stress.

GARDEN SECRETS

From Motubu, Craig, Greg, Rico, and I ventured back inland and headed north along the coastal highway toward

Oku. We'd heard about several centenarians living in remote villages on the island's northern extreme. On one side of the road, rain fell on mountains that sloped steeply upward, disappearing in the clouds; on the other, a leaden sky hung over the steel-blue Pacific. We passed the village of Ogimi, where Sayoko and I had met Ushi Okushima five years earlier. I made a note to stop and see her on the way back.

"What most impressed you guys about Kamada?" I asked, seeding a conversation. These car rides, I found, were excellent opportunities to tap into Craig's and Greg's expertise.

"Did you see what Kamada ate?" Greg began. I looked back at him from the front seat. His blue eyes widened as he spoke. He wore a short sleeve, plaid shirt and spoke with a slight, endearing lisp. "Some boiled vegetables and daikon, some carrots and miso soup. Maybe some stir-fried vegetables. All very simple, with nothing processed. People don't realize how bad sugar and meat are for them over time."

"Why's that?"

"Its called bio-ecological prevention, or risk. Let me explain. Simple, nonprocessed foods, often found in rural societies are associated with positive ecology of friendly bacteria in our intestines. These friendly bacteria include immunomodulating and fiber-fermenting lactic acid bacteria," Greg said.

"Stressors of this healthy ecological system," he continued, "such as surgery, certain medications, and consumption of meats and processed foods disrupt a natural balance and shift from friendly to unfriendly bacteria. This shift results in increased risk for diseases of urban or 'sophisticated' societies, such as inflammatory bowel disease, colon

"To be healthy enough to embrace my great-great-grandchild is bliss," says Kamada.

cancer, and more. It's associated with a low-grade systemic inflammation that is related to increased risk for all diseases of aging, like osteoporosis, heart disease, and dementia."

"Aren't these diseases occurring because people who eat junk food get fatter?"

"Obesity is certainly a risk factor," Greg replied. "But eating junk food also creates chronic inflammation of the gastrointestinal tract. The inflammatory response is good if we have an infection, but triggering it all the time by eating bad foods causes the body to produce chemicals that wreak havoc on our organs and arteries. People think that our skin is the main way our bodies interact with the outside world, but it is actually through our digestive tract—our stomach, large intestines, and small intestines. It has a surface area about the size of a tennis court. That's a lot of inflammation for our bodies to deal with."

"I noticed lots of mugwort," he continued. "Lots of turmeric, lots of garlic. Most people think mugwort is only for the Harry Potter books. Okinawan mugwort is one species of the genus *Artemisia,* which contains the most powerful natural substance for fighting malaria. The World Health Organization recently made it a top priority for one form of *Artemisia* to be available in developing countries."

"Here in Okinawa it is nearly a weed," Greg said. "It grows everywhere. People eat it all the time and use it for medicine, including the treatment of fevers. Did you notice all the turmeric? Turmeric is one-fifth as powerful as *cisplatin,* which is one of the most powerful drugs in chemotherapy. Turmeric is an anti-inflammatory, antioxidant, anticancer. This comes back to inflammation. Many age-related diseases are caused by an immune system out of balance. Excessive or unnecessary inflammation accelerates heart disease, bone loss, Alzheimer's disease. Antioxidants found in vegetables and herbs are also important, because the same oxidation process that rusts our cars also deteriorates our bodies. Antioxidants mop up highly charged oxygen-free radicals that cause this sort of rust.

"Kamada's garden where she gets her breakfast, lunch, or dinner is more than a natural grocery store. It's a pharmacy." Now, Greg was gesticulating expansively, making great circles with his hands. He had an endearing way of getting very excited about matters of health and complementary medicines.

"Okinawans see vegetables. I see powerful anti-inflammatory, antiviral, anticancer drugs," he said. "You know, you don't just wake up one day and have cancer. It's a process, not an event. And prevention is the same way; it has to be a daily activity. We're used to a world of options, and we think, 'Of all the options, what are the best?' Okinawans are just born into a lifestyle that promotes health. They have been blessed by access to year-round fresh, organic vegetables, strong social support, and these amazing herbs that amount to preventive medicines."

Craig had been sitting next to Greg, listening patiently. Now he jumped in. "It is not only *what* Okinawans eat, but *how much*. Their food has a very low caloric density, yet it is very nutritious. Compare a typical Okinawan meal—a tofu stir-fry, some miso soup, and some greens to an American hamburger. The Okinawan meal has three or four times as much volume and more nutrients but only about half the calories of a burger. So you're feeling fuller but getting leaner and living longer. And if you remember Kamada's hara hachi bu and eat only until you're 80 percent full, you'll even do better."

"How do you think this idea of hara hachi bu came about?" I asked. "After all, it's not natural to want to quit eating before you're full." I explained my theory of *cultural evolution*. After leading 17 expeditions to explore ancient mysteries, I noticed that over time, customs and traditions of successful cultures seem to undergo an unconscious but intelligent natural selection: Practices not good for a society tended to disappear while beneficial ones often survived—no matter how counterintuitive they may seem. I remember a story about a tribe in sub-Saharan Africa that cooked over open fires *inside* their huts. The huts filled with smoke that the villagers breathed. When a Peace Corps worker saw this he reasoned that the people's lungs were blackening with smoke. He asked them why they cooked indoors. When no one had an answer for him, he convinced them to move their cooking fires outside. Soon the people started contracting malaria at an alarming rate. It turns out that the smoke kept malaria-carrying mosquitoes out of the huts. This outweighed the negative health effects of the smoke.

Similarly, I don't think the first person who ever chewed a hot pepper thought, "Mmm, good." Capsaicin, pepper's active ingredient, is literally caustic to the flesh. But somehow, human taste has evolved to enjoy the taste of pepper. Why? Because capsaicin is a natural disinfectant, and it kills many types of food-borne bacteria. Put hot pepper in slightly rancid meat, and it inhibits bacteria. The person who eats the meal with the pepper lives. The person who eats the meal without the pepper gets sick and could die. Over time, the survivors acquire a taste for it, and a healthy culture evolves. "Can we say Okinawan longevity came with a similar cultural evolution?" I asked my traveling mates.

"It makes sense to me," replied Craig. "Mugwort tastes horrible; it is very bitter. But somehow, Okinawans have developed a taste for this bitter herb and commonly use it as seasoning in their rice. Is that because they have always liked it or because their taste has evolved with what is really good for them? The same thing has happened with the vegetable known as goya (bitter melon), come to think of it. Goya looks like a wart-covered cucumber. It's bitter as hell. Its juice is actually an astringent, the kind of food that makes you wince. But it contains high amounts of antioxidants and three compounds that lower blood sugar. Are Okinawans eating goya because it holds some kind of primal appeal to human taste buds? I don't think so."

TERRIFIC TURMERIC

Often encountered today in spicy curries and mustard, turmeric has a long history of use as a culinary spice and an herbal medicine; the Ayurvedic, Unani, and Siddha systems, as well as traditional Chinese medicine, advocated its use. Research has shown this spice's most active component, curcumin, has antioxidant properties and may help ease inflammatory ailments.

As we came into the northern extreme of the island, and the road narrowed through thick jungle. This, I thought, must have been what Okinawa looked like when centenarians were living the first halves of their lives.

We arrived at Oku in early evening. I passed a convenience store on the outskirts of a very small town and stopped to buy some snacks. As I paid for my wasabi peas, I asked the teenage girl in the school uniform if she knew anyone in the area who was over 100. I knew at least two centenarians lived in the area.

"Across the street," she replied. "Gozei Shinzato."

Greg, Craig, and I checked into a hotel down the block, knowing where our first call in the morning would be.

OUTER LIMITS OF THE HUMAN MACHINE

At 7 a.m. the following day, we walked from the road, crossed the bridge spanning a burbling stream, and passed through a garden to a simple, two-room dwelling on stilts. The rice paper screen was open, indicating that the occupant was up.

"Gozei!" I shouted.

We waited outside. It was raining, of course; a curtain of water fell from the roof and spattered on the mud at our feet. Moments later Gozei emerged from her bedroom. She was elfin, barely over four-and-a-half-feet tall, and wrapped in a kimono and overcoat. Her bare feet padded softly on tatami mats. She saw us, three huge American men wearing blaze-colored raincoats, and flinched. Then she burst out laughing—chortling actually—exhaling visible puffs of mirth into the cold morning air. Her tanned, wrinkled skin squeezed into a puffed-cheek, squinty-eyed smile.

"Come in," she said. We removed our shoes, stepped up into her house, and sat down. She warned us that she spoke the Okinawan dialect and only a little Japanese, so conversation might be difficult. Nevertheless, I managed, with a few dozen questions, to tease out her story.

For most of her life, Gozei had worked in the mountains barefoot, cutting firewood and carrying it back to the village to sell. When she was 18 her parents arranged a marriage to a local farmer. He had four children, whom she raised through two famines and many hard times.

Once when she was working in the mountains, she came upon a much bigger woman who had been bitten by a *habu*—a potentially deadly poisonous viper indigenous to the island. Gozei, who weighs about 85 pounds, cut off a strip of her dress to apply a tourniquet to the bite, then hoisted the woman onto her back. She carried the woman four-and-a-half miles back to sea and into a boat that she rowed to a neighboring village for help. The woman survived. Gozei was 62.

Now, 40 years later, Gozei still lived independently in a tiny wood-and-rice-paper house. Her husband died after 51 years of marriage. She tends a vegetable garden daily and harvests three times annually—mostly garlic, bitter melon (goya), scallions, and turmeric. She reads the comics her grandchildren give her and loves watching baseball games on television. Late afternoon is her favorite time of the day, when neighbors stop by for a visit.

As the conversation dwindled, we smiled nervously at each other. "Look at her face," Greg whispered. "That deep-skinned tan that comes from constant sun exposure. Did notice how Kamada's skin was the same? That probably means they've both had steady doses of vitamin D their whole lives."

"So?"

"I think vitamin D is an important ingredient in the longevity recipe," he said enthusiastically, as if just struck by an epiphany. "Your skin manufactures vitamin D when it comes into contact with the sun. Without that vitamin D, we increase our risk for nearly all age-related diseases including many types of cancer, high blood pressure, diabetes and even autoimmune diseases like MS (multiple sclerosis).

"Insufficient vitamin D markedly accelerates heart disease in kidney patients. For everyone, without enough vitamin D, bones become brittle, key hip and leg muscles become weak, and the chance of falling and breaking bones soars. When people this old break hips," Greg nodded toward Gozei, "they usually die very quickly."

"But I thought staying out in the sun too long *gave* you cancer, especially with all these changes in the ozone layer," I said.

"Sure, if you burn easily and don't get much sun in the first place," Greg rejoined, still whispering. "It's different now than it was for our parents—you don't want to go out and fry yourself. Burns are bad but staying inside being afraid of the sun isn't healthy either."

"Someone like Gozei," Greg continued, "who has a deep tan and is used to being outside all of the time, is optimizing her vitamin D. People who stay inside all the time would have to drink gallons of fortified milk every day to optimize their vitamin D. The amount of vitamin D in a multivitamin is not enough, especially if you are living in relatively northern cities such as New York or Chicago or have dark skin or work long hours or have celebrated many birthdays. Vitamin D in our bodies controls key elements of the immune system, blood pressure, and cell growth, and is important for cancer regulation. And in the test tube, vitamin D kills the cancers that most often kill Americans."

Gozei, who had waited patiently as Greg and I talked, hoisted herself to her feet and ambled into the next room. I waited a moment then followed her. She'd gone into the kitchen, a narrow room illuminated only by a small open window that admitted feeble light from the rainy day.

I stood quietly and observed. She reached under the sink, folded down a battered wooden box that served as a step, and climbed up to reach the faucet. She filled a teapot, pivoted to put it on the stove and turned on the burner. On the counter sat two clear jars, infusions of some sort—medicines I later learned. One of them contained garlic cloves; the other some green herb.

As she waited for the water to boil, she energetically washed a few dishes and, while she was at it, removed her dentures and scrubbed them too. When the pot whistled, she removed it from the stove, squatted in the corner, and poured boiling water over tea leaves. The room filled with the delicate, sweet aroma of jasmine tea. Her movements were slow and deliberate with a patient resolve. She seemed completely oblivious to me.

Back on the tatami mat, we sipped tea and made a futile attempt to restart our conversation. We were complete strangers from different hemispheres, indeed different centuries and were now at a loss for words. She smiled understandingly. Then she remembered the snacks! She stood up in one fluid motion and disappeared into the kitchen again. She emerged much later with two plates; one with lumps of raw sugar (a favorite of her grandchildren) and another of dried minnows (her favorite). She pinched off the head of one with a thumbnail and popped it into her mouth, instructing me to do likewise. The minnow was chewy, salty, and predictably fishy-tasting, but not unpleasantly so. I chased it with a sugar lump and a sip of tea. We smiled at each other some more.

Moments later she sprang up again. This time it was to make the daily offering to her ancestors, a ritual that is a cornerstone of Okinawa spiritual life. She stood in front of the far wall where a built-in shelf held a collection of vases with flowers, urns, and old photographs. Gozei lit a few sticks of incense and set them down in front of an ancient photograph of a sullen-looking peasant couple. A ribbon of smoke curled upward and filled the room with the smell of sandalwood. Again, I seemed to disappear. For the next ten minutes she recited a series of prayers bowing toward the altar. Then she sat back down and smiled.

"Do you see what's going on here?" Craig asked me. "This is what we call ancestor veneration. Older Okinawan women have great respect for their deceased ancestors. They believe that if they make the proper offerings in the

HOW MUCH SUN?

The body needs sunlight to manufacture vitamin D, but too much sun can damage the skin. To find balance, the National Institutes of Health advises 10–15 minutes of sun exposure twice weekly, then applying sunscreen with an SPF of 15 or higher afterward. This should provide enough sun to produce vitamin D without exposing the skin to too many damaging UV rays.

morning, the ancestors will watch over them for the rest of the day. It's like if something bad happens, it was meant to happen; if something good happens, it's because the ancestors were looking out for them. It's a great stress reducer for these people. They relinquish worries to a higher power."

"Cool," I said.

"Also," continued Craig, "do you see how she springs up and down from the mat?" How may 80-year-olds back home can get up from the floor like that? Gozei's over 100 and probably gets up and down 30 times a day. For her age, she has incredibly good lower body strength and balance. That makes a huge difference in old-age mortality because falls and broken bones are usually fatal when seniors reach a certain age."

At around 9 a.m., she excused herself to go make breakfast. Hoisting herself to her feet for the ninth time during our visit, she hobbled to the kitchen stove where she lit a flame under yesterday's soup. She spooned in fresh carrots, radishes, tofu, a tablespoon of miso paste, and let it heat. Meanwhile, she moved up and down the kitchen wiping clean the counters, sink, and even the window. Then she pulled up a chair facing the stove to wait for her soup. The flame cast a feeble light on Gozei's face. It occurred to me that I was witnessing the happy limits of the human machine. I sensed neither the frailty nor the wistfulness of impeding death but rather serenity—a certain satisfaction with a life now free of the ambition and commitments that dog younger years—a life achieved.

She poured her warmed soup into a bowl, gazed at it for a few long moments and murmured, "Hara hachi bu." She threw me a quick glance and then looked back at the steaming bowl, seemingly waiting for something. Then I realized: Perhaps she wanted to eat in private? I announced that I had to leave. "Thank you very much," I said to her, bowing slightly. "May I come back and visit again?"

"If you must," she quipped. And she chortled again, with her broad smile.

POWER OF NOW

Greg and Craig returned to Naha, but before they left, Craig gave me a lead on another location in the Okinawan Blue Zone: He told me to go back to Motobu and take the ferry to Ie Shima, a tiny island about ten miles off the mainland. There, his records showed, I would find 8 people over 100 living among a population of 5,300—an astounding number of centenarians. "Good luck, Danmeister. Tell me what you find."

That afternoon, Rico and I took the 30-minute ferry ride to Ie Shima and watched the tiny island, dominated by a solitary 551-foot-high volcano, come into view. Until now, we'd focused on Okinawa's emerging fame as a Blue Zone, but for many Americans, it has a darker reputation. The Battle of Okinawa was the largest amphibious assault of the World War II Pacific campaign. From April to June of 1945, the Allies mounted 1,300 ships and tens of thousands of troops against the Japanese. Some 70,000 American soldiers and Navy personnel were wounded or lost their lives; estimates for Okinawan civilian deaths range from 100,000 to 150,000.

We tend to lump Okinawans with the Japanese, but they're actually a distinct race—largely peasants—of the Ryukyu Kingdom subjugated by the Japanese in the late 19th century. They possessed none of the imperial ambitions of their Tokyo-based overlords, who forced them to fight in World War II. Centenarians recount how Japanese soldiers used their Okinawan draftees as human shields, forced to confront GI machine guns though armed only with bamboo spears. American warships rained down some 600,000 shells and fired more than 1.7 million rounds from the ground in the battle known as the "typhoon of steel." The assault literally changed the island's topography.

Ie Shima's part in the Battle of Okinawa lasted six days. Many Okinawans were killed. But one of the dead was Ernie Pyle, the Pulitzer Prize-winning war correspondent whose dispatches captured the heroism and humanity of the American G.I. during World War II. His heartbroken comrades erected a wooden cross marking the place where he died from a bullet wound to the head. Today an inscription on a stone monument reads: "At this spot, the 77th Infantry Division lost a Buddy, Ernie Pyle, 18 April 1945."

When Rico and I landed, we rented bicycles and peddled to city hall to make an appointment with the mayor. We needed to ask permission to set up interviews with centenarians. We fully expected an overnight wait for a meeting. Instead, an assistant brought us right in to see the mayor in his large, fan-cooled office and offered us green tea.

"How can I help you?" the mayor asked.

An affable man of perhaps 35, he seemed removed from the crushing protocol of Naha. We asked him if he knew any of the eight centenarians that we had heard were living on the island.

"I know one," he told us. "Kamata Arashino. She's an incredible lady. Everybody knows her as a local hero." He went to a file cabinet and pulled out a booklet of local history, then thumbed through to the page about Kamata and began reading. Rico translated:

> It was a rainy day in April. Forty-three-year-old Kamata Arashino and her 3 children were hiding with 130 other villagers in a cramped cave. Five days earlier, American troops had stormed Ie-jima and had killed nearly half of our island's population. Battleships were shelling the island from offshore. The only hope for these poor peasants was to take shelter in a cave. They were told that if American soldiers captured them, they'd be tortured to death. So in the event of capture, villagers were given suicide bombs they could detonate for a painless ending. After several days, American troops landed on western shores of the island. Troops advanced from the beach toward the cave. The villagers made a hasty decision to detonate the bomb. But a split second before the bomb went off, Kamata decided that she wanted to live. She rushed her children to the back of the cave. There was a white flash, an ear-shattering blast, and the cave's roof collapsed....

"That was an amazing story," I said. "And is Kamata still alive?"

"Yes," the mayor said to me. "She still lives right here in the village."

"Can I see her?"

"Why, yes." The mayor led us outside city hall and pointed down the street to a simple house close by. It was only about a block away.

Seventy-four-year old Shigeichi, a bald man with a permanent smile opened the door. "Kamata is my mother," he said when I told him I was looking for the woman who had survived the cave blast. "I was in that cave with her." He invited us into his living room, where we took off our shoes and sat cross-legged at a low table. He poured us green tea.

He laughed good-heartedly at my questions about his life and that of his mother. "Our life was very difficult before the war," he began. "Like most people here, we had to live off of what we produced. We ate sweet potatoes three meals a day, and maybe some fish. Once a year, during the lunar New Year, we'd slaughter the family pig and eat pork. Mostly we were hungry, though." He chuckled.

"Then the war came. There was no food. We ate miso and drank rainwater. When the American troops landed here in April 1945, they ran up the beach firing machine gun—tot, tot, tot," Shigeichi mimicked machine gun fire. "Battleships shelled us from the sea. Over half of the people on the island died."

"Does it feel strange to have an American in your house?" I asked, feeling mildly guilty for representing "the other side" of that war.

"No." He dismissed the question with a wave. "That was a different time." Then I asked him about the cave and how the mayor had told us his mother's story.

Okinawans maintain strong social connections through regular gatherings of a moai.

"It was misery," he said, the smile leaving his face. "We were hiding from American bullets. We had no food, little water; it was so crowded you could barely lie down. Then somebody shouted that the Americans were coming. My mother is a strong woman. She wasn't going to let us die. Just before the bomb went off, she threw us to the ground and put a futon over us. I heard a bang so loud that my ears still ring today." He tapped his hearing aid. "Big chunks of the ceiling fell on us."

"What I most remember was the terrible silence after the bomb went off. Over 110 people died. Miraculously, my family lived.

"The Americans captured us but did not hurt us. Instead, they fed us rations. Eventually, the war ended and prosperity came. My mother lives with my wife and children. Every afternoon after school, the grandchildren visit with their grandmother."

Months later on a return trip to Okinawa, I was to meet Kamata in the old-age recreation center where she spends her days. Inside, 30 or so seniors sat at tables chatting or doing crafts. It looked like a bright, cheerful kindergarten with hand-drawn pictures on the walls and board games on the table. But it smelled, as these places often do, of urine. Bent old men hobbled by with walkers; one lady slouched in a chair with a string of drool hanging from her mouth like clear spaghetti. Though Kamada and Gozei don't have access to this type of recreation, it struck me that they were better off with their gardens, their moais, and the weekend visits from grandkids.

We found Kamata sitting off to the side folding towels and chatting with three other women about her age. She was small—no taller than four feet—and dressed in a color-ful shirt. She had short, white hair that she combed straight back from her forehead. Her century-old face was wrinkled like a pumpkin after a hard freeze. She was nearly deaf, but she was sharp, quick-witted. Her eyes darted about in a way that showed she was aware of everything going on around her. She smiled when I sat down next to her.

I told her that I'd traveled the world interviewing cente-narians and that her story was the best one that I had heard. Could I ask her about it?

"I endured great misery," she said, striking a chord com-mon to many Okinawan centenarians. "I was hungry all the time for many years. My husband and my first son were killed in the war."

I asked her about the cave. "Yes, I was there. There was a great explosion and I lived, my children and I." She looked up and saw me taking notes. "This is enough!" she said, chopping the air with her hand. "I'm tired of the past. I don't want to talk about it. I'm happy now. I have enough to eat. I'm surrounded by my friends. Why relive misery when better times have arrived? I've lived those hardships, and now they serve me well because they allow me to enjoy today."

LIFE: A POPULARITY CONTEST AFTER ALL?

Was Kamata's ability to put the past behind her and live in the moment one of the stress-coping mechanisms that explained her longevity? Or was it the hardship, meager diet, and sense of belonging that came with living with family—and being there for grandchildren? With few exceptions, the healthiest centenarians possessed lifestyles and temperaments very similar to those of Kamada, Gozei, and Kamata.

Over the next three weeks, Rico and I were to meet dozens of centenarians—all different variations of each other. We interviewed Yoshiei Shiroma, a tanned, leather-skinned man from Naha, who told us that candy was the secret to his longevity. He not only ate it every day, he also made his own, crushing sugar cane and boiling it down in his backyard.

"If you eat so much candy," I asked him, "how come you have such nice teeth?"

"They're false," he laughed, tapping his white dentures.

Sitting on the floor was part of the daily routine for 94-year-old Koutoku Kinjo. It helped keep him flexible so that he could still practice *bojutsu* (a martial art using a stick) and ride his motorcycle to his garden every day.

Fumi Chinen has never used a dirty word—not once—in all her 99 years. We met her at her clothing stall in the Naha market. With her sweet smile and hair pulled back in a bun, she looked like the kind of grandma everyone would want. Her advice for longevity included: "Eat eel every day, work in a place where you can socialize, and if anyone ever gives you something and tells you it's healthy, don't eat it!"

To stay young, innkeeper Ishikichi Takana prays every day. "My ancestors are watching over me," the 99-year-old told me. "I never pray for a long life, but I just express my gratitude for another day. It reminds me that every day is important." We visited Seiryu Toguchi, a 105-year-old man whose ikigai was tending his garden and playing his lute. His neighbors all looked after him out of affection.

I spent a morning on a Naha beach working out with Fumiyasu Yamakawa, a one-time banker. Every day at 4:30 a.m., he cycled to the beach, swam a half hour, ran a half hour, did yoga, and then met with a group of other Okinawan seniors who stood in a circle and laughed.

"Why is that?" I asked.

"It's vitamin S," he said. "You smile in the morning and it fortifies you all day long."

ALL DIFFERENT

The common attributes of these Okinawan centenarians—their hard-edged humor, affably smug dispositions, hardship-tempered appreciation for what is, not what could have been, and their purpose-driven, tradition-based life-style—might look like clues to their longevity. But they wouldn't hold much scientific sway. You can't generalize about a whole population from a few stories. That's where Dr. Nobuyoshi Hirose, one of Japan's foremost centenarian researchers, came in.

Earlier during my stay in Tokyo, Greg had set up a dinner with Dr. Hirose. A short, amiable scholar with an odd sense of humor and a wheezing laugh, Hirose described 15 years of centenarian research. During the past 40 years, he said, Japan's centenarian population had mushroomed. I asked him if he could explain why.

"The only common factor we could find is the heterogeneity of centenarians," he said. "In other words, they are all different."

Like all credible longevity experts, Hirose avoided definitive conclusions. But as the night wore on, and the sake flowed, he loosened up. Hirose had discovered that the daily intake of protein, fat, carbohydrates, and total calories was lowest among centenarians (mostly because of their lower body weight). The big difference between centenarians and younger populations was the greater appetite for vegetables, and especially for dairy products, among centenarians. They didn't necessarily eat more or less than others, relative to body weight, but what they did eat was rich in calcium, vitamins, and iron.

Hirose reached down for his briefcase and pulled out a map with tiny red dots to illustrate the distribution of centenarians in Japan, each dot representing ten centenarians. In the northernmost islands, there was a sparse scattering of dots. But the dots thickened toward the south, and in Okinawa, the cluster was so dense that it appeared as a red blotch on the map.

"Japan has seven semi-supercentenarians per million people," Hirose said, referring to those over age 110. In Okinawa, the rate is 35 semi-supercentenarians per million. "Perhaps it's because in the north, where it's colder, old people are more likely to die of respiratory infections. Or because in Okinawa they can grow vegetables year-round and therefore eat fewer salty pickles and canned meat. Or perhaps exposure to more sun gives Okinawans an advantage?" That would support Greg's vitamin D theory.

> ## "SUPER"
>
> A nonagenarian is a person who is in his or her 90s, a centenarian is someone who reaches the age 100 and older, but a supercentenarian is a person aged 110 and higher. Today, the exact number of supercentenarians in the world is unknown, but studies show that worldwide their overall numbers have been increasing steadily since the 1980s.

He shifted gears to describe the work he and his researchers had been doing on happiness.

"We see that people over 40 tend to get less happy with age, until about age 80," he said, drawing a U-shaped curve on a piece of paper. "That's when their well-being begins to curve upward again. By the time a woman reaches 100, she is happier than a 40-year-old, even though she's probably functioning poorly. This is due to our favorable social environment. Americans emphasize *biological aging*. You tend to age alone. In Japan we focus on *social, environmental aging*. We think about aging in the context of a family or community."

Centenarians also tend to be decisive, he continued. "They know what they want and then stay on course. But when life circumstances force them to adapt, they become flexible thinkers, able to embrace the change. Almost always, they are very likable people too." Even those who may have been cantankerous in their youth learn the value of good humor and grace in cultivating the loyalty and patience of their friends and caregivers as their level of functioning declines. They make it fun and rewarding to be around them."

ANOTHER VISIT WITH USHI

Before I left Okinawa, I paid one final visit to 104-year-old Ushi—Okinawa's emblem of its Blue Zone. After three weeks of rain, it was a wet, gloomy afternoon when I arrived in Ogimi. I spent some time wandering through the patchwork labyrinth of garden-and-hut-lined streets. People were inside, I imagined, sitting on the floor, drinking green tea, or napping warmly in their bedrolls. From one low, wooden house with a dripping, slanting roof I heard the ringing of laughter—big, hee-hawing, belly laughing. Ushi.

Inside her house, Ushi sat snugly wrapped in a kimono. Her hair was combed back revealing her bronzed forehead and alert, green eyes. Her smooth hands lay serenely folded in her lap. At her feet, her 77-year-old daughter, Kikue, and best friends, 90-year-old Setzu Taira and 96-year-old Matsu, sat cross-legged on tatami mats. Kikue, who I later learned was very protective of her mother, dutifully poured me a green tea. I began to ask questions.

Since I'd visited Ushi with Sayoko, Ushi had taken her first paying job, tried to run away from home, and had begun wearing perfume.

"Perfume?" I asked.

"She has a new boyfriend," Setzu offered. "He's only 75 years old."

I look at Ushi as she clapped a hand over her mouth and unleashed one of her blessed hee-haw laughs. When the echoes faded, I sat quietly and let the ladies resume their conversation.

The house looked the same as it did during my last visit: the low table where Ushi takes meals, the ancestor shrine in the main room, the simple futon roll visible through the open bedroom door, the age-browned wooden wall, the tatami mats worn smooth with decades of use. Near a back door I saw soiled gloves, wooden thongs, and the large conical hat—Ushi's garden wear. For the next few minutes, I was happy to immerse myself in the exotic simplicity of hot jasmine tea on a cold day, sheltered from the rain in the simple conviviality of rural Okinawa.

I looked over at Setzu, who was sitting cross-legged next to me. Gray strands streaked Setzu's otherwise inky-black hair. I was sitting several feet away from her, but could smell the fields she'd been working that morning, the scent of perspiration and earth. With a gentle hand on Ushi's arm, she drew herself close to her old friend's ear to recount some memory. (Perhaps of my last visit?) After a moment, she noticed me staring at her, and with a side glance, smiled at me. But then she quickly turned away, shyly. No doubt she remembered the secret she had told me five years earlier.

The first time I met Ushi and Setzu, we'd had lunch together. I was here for the 2000 "IslandQuest" interactive expedition. Over a meal of stir-fried bitter melon, kelp, millet rice, and fennel tea, I had interviewed them about their longevity. The conversation started off predictably enough, with me asking questions of the what's-your-secret variety. They answered dutifully. Each of them had come from a life of hardship. As children, they nearly starved. During the war, they'd hidden in the mountains and lived on berries. Setzu remembers being caught by an American soldier while she was searching for food. The soldier told her to get into a line with other Okinawans. She thought she would be shot. Instead, the Americans handed out chocolate and biscuits to them.

"When I tasted that chocolate," she'd said, her eyes welling up, "I knew my children would live." She'd looked down and sobbed silently, then looked up again. "I've waited all my life to thank someone for that. So now I thank you."

Now, seeing her again at Ushi's house, I remembered feeling a strong sense of embarrassment and pride at hearing Setzu's story. As an American, I was loosely connected to both the cause of her suffering and her salvation. I also knew I was inexorably linked to the beginning of the end of her way of life.

In prewar Okinawa, hardship had tempered the majority of people who lived to become centenarians: periods of hunger, of discipline, of physical exertion, of eating bitter-tasting but healthy foods, like goya. When America won the war and established a military base on Okinawa, it brought peace, prosperity, jobs, and a culture of rich fast food and huge portions. But as is the case with most stories of development, prosperity arrived as a paradox: The end of economic hardship also brought an end to the same disciplines, lifestyles, work requirements, and the diet that had helped foster the culture's extraordinary longevity.

After being starved for so many centuries, Okinawans seized this new food culture. They quickly developed a taste for canned meat (Hormel to this day still exports approximately five million pounds of SPAM a year to Okinawa) and fast food (Okinawans eat more hamburgers per capita than any of Japan's other 47 prefectures). A sharp increase in obesity-related diseases such as diabetes has ensued. Okinawa now has Japan's highest rate of obesity, and, among middle-aged men, one of the highest rates of premature deaths from cardiovascular diseases.

While Okinawan women are still among the longest-lived people on earth, men are dragging down the region's average life expectancy. Formerly the longevity leaders (and still the leaders in remaining life expectancy at age 65), by the year 2000, Okinawan men had dropped to the middle of the pack, sitting in 26th place among 47 prefectures in life expectancy at birth. In other words, we can now witness two distinct groups in Okinawa: A healthy, older generation with a longer life expectancy, and a distinctly less healthy younger generation whose life expectancy growth has slowed such that they are losing ground in comparison to other Japanese prefectures. One of the main culprits: a brand of prosperity imported from the United States.

I sat with Setzu, Matsu, Ushi, and her daughter for another hour. Outside, dusk was falling. Ushi's daughter shot me a glare that meant, "You've overstayed your welcome." Ushi,

Matsu, and Setzu seemed to take the cue and fall silent in unison. These women belonged to the same moai for nearly a century, and now they seemed to communicate wordlessly. I had one more question.

What was Ushi's ikigai, her pervasive sense of purpose? "It's her longevity itself," blurted Ushi's daughter. "She brings pride to our family and this village, and now feels she must keep living even though she is often tired."

I looked at Ushi for her own answer.

"My ikigai is right here," she says after a moment with a slow sweeping gesture that takes in Setzu, Matsu, and her simple surroundings. "If this goes away, I will wonder why I'm still living."

OKINAWA'S LONGEVITY LESSONS

Among the centenarians in Okinawa's Blue Zone, these practices are common.

Embrace an ikigai.
Older Okinawans can readily articulate the reason they get up in the morning. Their purpose-imbued lives gives them clear roles of responsibility and feelings of being needed well into their 100s.

Rely on a plant-based diet.
Older Okinawans have eaten a plant-based diet most of their lives. Their meals of stir-fried vegetables, sweet potatoes, and tofu are high in nutrients and low in calories. Goya, with its antioxidants and compounds that lower blood sugar, is of particular interest. While centenarian Okinawans do eat some pork, it is traditionally reserved only for infrequent ceremonial occasions and taken only in small amounts.

Get gardening.
Almost all Okinawan centenarians grow or once grew a garden. It's a source of daily physical activity that exercises the body with a wide

range of motion and helps reduce stress. It's also a near-constant source of fresh vegetables.

Eat more soy.

The Okinawan diet is rich foods made with soy, like tofu and miso soup. Flavonoids in tofu may help protect the hearts and guard against breast cancer. Fermented soy foods contribute to a healthy intestinal ecology and offer even better nutritional benefits.

Maintain a moai.

The Okinawan tradition of forming a moai provides secure social networks. These safety nets lend financial and emotional support in times of need and give all of their members the stress-shedding security of knowing that there is always someone there for them.

Enjoy the sunshine.

Vitamin D, produced by the body when it's exposed on a regular basis to sunlight, promotes stronger bones and healthier bodies. Spending time outside each day allows even senior Okinawans to have optimal vitamin D levels year-round.

Stay active.

Older Okinawans are active walkers and gardeners. The Okinawan household has very little furniture; residents take meals and relax sitting on tatami mats on the floor. The fact that old people get up and down off the floor several dozen times daily builds lower body strength and balance, which help protect against dangerous falls.

Plant a medicinal garden.

Mugwort, ginger, and turmeric are all staples of an Okinawan garden, and all have proven medicinal qualities. By consuming these every day, Okinawans may be protecting themselves against illness.

Have an attitude.

A hardship-tempered attitude has endowed Okinawans with an affable smugness. They're able to let their difficult early years remain in the past while they enjoy today's simple pleasures. They've learned to be likable and to keep younger people in their company well into their old age.

An American Blue Zone

An American Blue Zone

The Longevity Oasis in Southern California

MARGE JETTON BARRELED DOWN THE San Bernardino Freeway in her rootbeer-colored Cadillac Seville. Peering from behind dark sunshades, her head barely cleared the steering wheel. She was late for one of the several volunteer commitments she had that day, and she calmly but firmly goosed the Caddy's throttle to move it along.

It was early on a Friday morning, and Marge had already accomplished quite a lot. She had walked a mile, lifted weights, and eaten her oatmeal breakfast. "I don't know why God gave me the privilege of living so long," she said, pointing to herself. "But look what he did!"

Marge, born September 29, 1904, is one of some 9,000 Seventh-day Adventists who live in and around Loma

Linda, California, 60 miles east of Los Angeles. For the past half century, members of this community, whose faith endorses healthy living, have participated in a ground-breaking health and dietary study of Californians over the age of 25. The results of this study hold promising clues to another remarkable fact about these Adventists: As a group they currently lead the nation in longest life expectancy.

Marge Jetton, I decided, was the Adventist poster girl. She had sucked me into the whirlwind of her 100-year-old orbit an hour earlier at the Plaza Place hair salon just outside of Loma Linda. For the past 20 years, Marge has kept an 8 a.m. Friday appointment with stylist Barbara Miller.

"You're late!" Marge shouted as I burst through the door at 8:25. She'd been flipping through a copy of *Reader's Digest* as Barbara uncurled the bluish-white locks of hair that now dovetailed around Marge's head like a small cumulus cloud. Behind Marge, a line of stylists coifed hair on other ancient noggins. "We're a bunch of dinosaurs around here," Barbara whispered to me.

"You may be," Marge shot back. "Not me."

Half an hour later, Marge led me to her car. She didn't walk, quite, but scooted with a snappy, can-do shuffle.

"Get in," she ordered. "You can help."

The mission: Deliver recyclable bottles to a woman on welfare who will later redeem them for deposits. But first we hopped on the freeway and drove to the Loma Linda adult services center, an activity center for seniors, most of whom are several decades younger than Marge. She popped open her trunk and heaved out four bundles of magazines she'd collected during the week.

"The old folks here like to read them and cut out the pictures for crafts," Marge explained.

Old folks?

ISLAND IN THE BIG CITY

Despite its location in the smoggy orbit of greater Los Angeles, Loma Linda appears to be one of the few places in the United States where a true Blue Zone has taken root. Adventists like Marge Jetton follow a faith that expressly discourages smoking, alcohol consumption, or eating foods deemed to be unclean in the Bible, such as pork. In fact, the religion discourages the consumption of meat in general, as well as rich foods, caffeinated drinks, and even "stimulating" condiments and spices.

Some of the most conservative Adventists don't believe in going to the movies or the theater or indulging in any other form of popular culture. All of these tenets seem to have helped turn Loma Linda into a longevity oasis right in the middle of America's second largest city.

Driving to America's Blue Zone from Los Angeles International Airport was a classic California freeway experience. As I sped along in my lane no more than a foot or two from vehicles on either side, I felt as if the road itself was propelling us along. Looking at the brownish-yellow air draped over the mountains beyond the strip malls, it wasn't hard to figure out where all the exhaust was settling. When I took the Anderson Street exit for Loma Linda, I passed a smorgasbord of fast-food franchises all fighting for attention. Could this *really* be the doorstep to an American Blue Zone?

After just a half mile of steadily climbing up the road (Loma Linda is Spanish for "lovely hill"), I found myself surrounded by the manicured lawns, parking lots, and large buildings of the Loma Linda University and Medical Center (LLUMC). Privately owned and operated by the Seventh-day Adventist Church, and founded in 1905, LLUMC is a modern facility with a proud tradition of service. I knew that physicians at the medical center had played key roles in the Adventist Health Study and had collected a wealth of data on Adventist behaviors.

Julie Smith, LLUMC's director of public and media relations, took me on a quick tour, starting on the roof of the medical center where a pair of busy helicopter pads serves the only Level 1 trauma center in the area. In the basement of the Children's Hospital Pavilion, the world's first hospital-based proton radiation treatment system treats about 160 patients five days a week (many for brain or prostate cancer) and provides research for NASA scientists. Down the hall is a wall covered with pictures of happy, healthy children. Through the pioneering work of Dr. Leonard Bailey, a few hundred of the United States' infant heart transplants have occurred at LLUMC. The photographic display is a montage of survivors, including a strapping young man in his early twenties who was the world's first successful infant-to-infant heart transplant recipient in 1985.

After the tour, I made a long trek through the courtyard to a far corner of the campus. There I found a side entrance to Evans Hall, a building that once housed unassuming classrooms for medical and dental students and now housed the university's Center for Health Promotion.

The stairs under the frayed carpet creaked and moaned as I made my way to the second floor offices of Dr. Gary Fraser and Dr. Terry Butler.

As soon as the introductions were made, Dr. Butler asked me if I'd like a drink. Since it was a hot day, I said yes, imagining a cold soda or iced tea. Butler drew me a glass of room-temperature water from a large jug on his desk. More than a gesture of hospitality, this act was Butler's way of practicing what he preaches—and he and Fraser had shelves full of data to buttress their dietary message.

When LLU researchers first embarked on a study of the dietary habits of nearly 25,000 Adventists in California a half century ago, it was good news to the American Cancer Society, which had just initiated its own study of the effects of tobacco smoke on lung cancer. Because the overwhelming majority of Adventists were nonsmokers, they provided an ideal control group, and they were promptly folded into the ACS study.

As we now would expect, the subsequent data (collected from 1958–1966) revealed that Adventists contracted lung cancer at a rate of only 21 percent of that of the original ACS group. But what caught even early anti-smoking advocates offguard was that the Adventists also had a much lower incidence of other cancers, as well as less heart disease and diabetes. Even when compared to nonsmokers in the original ACS group, the Adventists produced generally healthier outcomes.

The results were intriguing. If the Adventists had created a longevity culture, was it possible to determine which of their dietary and lifestyle behaviors was most responsible for it? That was the thrust behind what became known as the Adventist Health Study-1 (AHS-1), a survey funded by the National Institutes of Health that examined nearly 34,000 California Adventists over age 25 from 1974–1988.

Dr. Fraser came to LLU to teach medicine and epidemiology in 1979, shortly after AHS-1 was initiated. Since 1987 he has been the principal investigator of the Adventist Health Study projects. (A second version, AHS-2, began gathering data on 97,000 Adventists in 2002 and will start reporting results in the next year or two.) A slim man with a precise manner, he is frequently asked to sit on panels that review grant proposals for other NIH studies.

Born and raised in New Zealand, he started out as a cardiologist, an endeavor he describes as "seeing people who after 40 or 50 years of not taking care of their bodies were reaping the results. It was like trying to close the door to the barn after the horse has bolted—very frustrating. When I discovered I was very good at math," he continued, "I decided to become an epidemiologist and see if I could help ward off heart disease on the front end, which is much more satisfying." His coinvestigator, Dr. Terry Butler, is an Adventist pastor as well as an epidemiologist. Fraser credits him with helping to stimulate high levels of Adventist participation in the AHS studies.

"There are a lot of things in AHS-1 to hang our hats on," Fraser said with a prominent down-under accent. "First, we can say with certainty that Adventists live longer." In California, the study showed, a 30-year-old Adventist male lives 7.3 years longer than the average 30-year-old white Californian male. A 30-year-old Adventist female lives 4.4

years longer than the average 30-year-old Californian white female. "If you go to Adventists who are vegetarian," said Fraser, "it becomes 9.5 years longer for men and 6.1 years longer for women. It is not surprising why this is so. About two-thirds of people either die of heart disease or cancer, and the Adventists do a number of things to protect themselves from heart disease and different cancers."

LOMA LINDA FACTS

- Nation: United States of America
- Location: Southern California, about 60 miles east of Los Angeles
- Loma Linda Population: ca 21,000 people

One of the key discoveries of the AHS-1 survey was that approximately half of the Adventists were vegetarians or rarely ate meat, which gave Fraser and Butler a solid demographic foundation to look at the advantages of a plant-based diet. "We learned that nonvegetarian Adventists had about twice the risk of heart disease as vegetarian Adventists," Fraser said, "particularly men but also younger and middle-aged women."

Back in the late 1980s and early 1990s, that finding was relatively controversial. "It didn't automatically mean that meat did the damage by itself, so we dug deeper," Fraser said. "But meat remained a consistent contributor to heart disease, which isn't that surprising, because it has a high level of saturated fat. So out of that we started asking questions about other fatty foods, and one of the ones we focused on was nuts."

At the time, clinical nutritionists were telling people to stay away from all fatty foods and snack foods, including nuts. "But," Fraser said, cleaving the air with both palms for emphasis, "it turns out that most of the fat in nuts is *unsaturated* fat. And when we looked at that data, it was really so clear: The Adventists who consumed nuts at least five times a week had about half the risk of heart disease of those who didn't. This was true of men, women, vegetarian, nonvegetarian—we split the population up about 16 or 17 different ways and each time asked the question, 'Does nut consumption matter?' And every time we saw that it did."

Since their findings were published in 1992, at least four major studies have confirmed that eating nuts has an effect. The American Heart Association has a positive recommendation about nuts. "So now everybody's asking, what is it about nuts?" Fraser said. "I'm not sure we have all the answers, but one thing certainly is that they have an effect on lower blood cholesterol."

"What about nuts that are roasted in oil?" I asked, glancing quickly behind Butler to see if there's a bag of cashews or almonds beside his jug of water.

"Doesn't matter," Fraser replied. "Nuts have hard, thin skin, so that doesn't have much impact."

"Now when you talk about AHS-1 and cancer, it gets a little more controversial," he cautioned. "Because, despite hundreds of studies and huge amounts of press, what epidemiologists know *with certainty* about diet and cancer can be stated in a single paragraph. And that would say that consuming fruits and vegetables and whole grains seems to be protective for a wide variety of cancers."

AHS-1 was one of the first studies to really demonstrate that, Butler added. "But exactly *which* fruits and vegetables and *how* protective they are with *which* cancers is hard to say with certainty, although we have some very interesting data. For example, we found that women who consumed tomatoes at least three or four times a week reduced their chances of getting ovarian cancer by 70 percent over those who ate tomatoes less often. Something like that gives you pretty good evidence that there is protection, but because of our limited sample size, knowing the degree of protection may be more up for grabs. Eating a lot of tomatoes also seemed to have an effect on reducing prostate cancer for men."

"Colon cancer also," Fraser said. "We found that the Adventists who ate meat had a 65 percent increased risk of it compared to the vegetarian Adventists. And Adventists who ate more legumes like peas and beans had a 30 to 40 percent reduction in colon cancer."

For pancreatic cancer there was again a small sample size, but researchers did see that those who ate fruits and legumes had a much lower risk. "Those who ate meat were at twice the risk of getting bladder cancer and a 65 percent increase in the risk of getting ovarian cancer," Fraser said.

"For lung cancer, we found an extensive relationship with smoking, which is no surprise, although most Adventists are past smokers if they smoked at all. We naturally have a large percentage of Adventists who never smoked, and when we looked at them, the ones who ate two or more servings of fruit per day had about 70 percent fewer lung cancers than those nonsmokers who ate fruit only once or twice a week."

"Tell him about water," Butler prompted.

"Right," Fraser said. "My personal view is that this is

a potentially very interesting—although overall still tentative—finding. But if you look at the AHS-1 data, it is very clear that men who drank five or six glasses of water a day had a substantial reduction in the risk of a fatal heart attack—60, 70 percent less—compared to those who drank considerably less water. The difference for women hasn't been seen as much. Yet if you looked at drinking nonwater fluids, consuming a lot of soft drinks, coffee, cocoa, seemed to be hazardous and increased the risk of attack. Imagine if you could really make a sizable dent in heart attack rates just by increasing the number of glasses of water you drink! That would be a major public health finding."

Fraser stared into his lap for a moment and pursed his lips. "For whatever reason, people just aren't looking into this," he said simply. "But we are looking at it again with AHS-2, and with 97,000 people, we should find it much more strongly if it is real. If it is proven again, it is huge for public health."

He grabbed his glass and offered a toast. "I'll tell you this: As a result of what I've seen, I make sure *I* drink five or six glasses of water a day."

BENEFITS OF A LIFESTYLE

So what were the key Blue Zone lessons about longevity from the Adventist studies? Fraser and Butler listed five things we can do to add as much as an extra decade to our lives.

"First, vegetarian status will get you about two years," Fraser said. "Not eating meat is clearly important, and our studies have shown us it is because it seems to have an impact on heart disease and some cancers.

"Second, we found that nut eaters also had a two-year advantage, which seemed to relate largely to heart disease. Of course there are causes of death not related to cancer and heart disease, and we suspect some of these behaviors might also be protective for some of those causes."

"Third is being a smoker," added Fraser. "Or even a past smoker, as we found among the Adventists. If you have ever been a smoker, it has a moderately strong impact on lung cancer and some impact on heart disease."

"Fourth is physical activity," he continued, "which again accounted for an extra couple of years, and that seems to run very clearly to heart disease and to certain cancers like breast and colon cancer. The evidence is fairly clear that most of this benefit comes from modest but regular physical activity. It really flattens out once you get to the marathoner level, which is not necessary for longevity."

The fifth and final recommendation is to maintain a fairly normal body weight, an increasingly important issue in the United States. "It turns out a lot of the very active chemicals that relate to inflammation probably have some impact on increasing the likelihood of cancer developing," Fraser said. "Those chemicals could well come from fat cells. One huge advantage of a vegetarian diet is how much it is associated with lower body weight. In fact we just cracked some numbers from AHS-2 that show that Adventists who are what we call lacto-ovo vegetarians, meaning they eat eggs and other dairy products, still are an average of 16 pounds lighter than Adventists of the same height who are nonvegetarian. And Adventists who are strictly vegan, which is only 4 percent, are 30 to 32 pounds lighter than nonvegetarian Adventists of the same height. That has a huge impact on cardiovascular disease, on blood pressure, on blood cholesterol, on inflammation related to hormones and the way it stimulates cells in the body."

I noted with a pang of regret that my nonvegetarian body was starving and could also still use a shot of caffeine. I drained my second glass of water, hoping that one of these good gentlemen would break out the nuts. But Fraser wasn't quite finished. "If I could stress just one point to you, it would be this," he said. "The Adventist experience, or lifestyle, we are studying doesn't have to be that unique. In many, many ways they are fairly typical Americans. I think the most exciting thing about our results is that even with modest changes there is no reason why non-Adventists can't benefit from these things in the same way."

GENESIS OF THE ADVENTIST WAY

Directly across from the hospital is the Loma Linda Market, as large as most national chain grocery stores but stocked with soy "beef jerky," egg-free ice cream, and more than 80 bins of various seeds and nuts. Lest you think this is an anomaly or the sole epicenter for the health food crowd, about a mile away is Clark's Nutrition Center, another supermarket-sized outlet for organic foods and supplements. One entire wall at Clark's is devoted to gluten-free nourishment. Another aisle contains huge, 50-pound bags of organic carrots, and coffee-table-sized flats of organic wheatgrass that could be mistaken for miniature lawn plots. And everywhere you look there are nuts: showcased in plastic-sealed gift boxes; in heaping piles beside the onions and potatoes in the produce section; and sorted—raw cashews, dried pistachios, honey-roasted almonds, yogurt-dipped peanuts, shelled and unshelled walnuts—in dozens and dozens of hanging bins that can be emptied into flask-sized paper bags with the pull of a lever.

Most of us have the best intentions of getting more exercise or eating more nutritiously. But religion has provided Adventists with the extra nudge that seems crucial for turning intentions into habits. We've all heard the phrase, "Cleanliness is next to Godliness." For Adventists, healthiness is next to Godliness.

"Being healthy has always been a fundamental part of the Adventist message," said Dr. Daniel Giang, vice president for medical administration at LLUMC. I met Giang in a spacious conference room in the Chan Shun Pavilion. "Maybe it's because I grew up in Takoma Park, Maryland, which is one of the world headquarters of the Adventists, but I've always been very interested in the history of our church, which is really fascinating when it comes to health," he said. An amiable man of Chinese-American ancestry, Giang was both enthusiastic and easygoing as he sketched out the Adventist story.

"I think most Adventists, especially if they grew up in the 1950s and 60s, think the church was formed simply because God spoke to Ellen G. White and she wrote down what He said," Giang told me. "But if you go and read the history, the early Adventists said there was a lot of hammering things out as the church was being formed in the 1840s, 1850s, and 1860s, with Ellen White being influential in saying whether people were on the right track or the wrong track," he said. That was especially true of the period between 1844, when White had her first vision from God, and 1863, when the Adventist Church was officially organized.

"Along with Ellen G. White, there were two people crucial to the way the church was formed. One was James White, Ellen's husband, who provided the administrative genius and ran a publishing company. The other major figure was Joseph Bates, a very successful sea captain and intellectual who introduced this whole health emphasis," Giang said. "Long before it became a part of the Adventist religion, Bates had decided to give up smoking, alcohol, and consuming any meat, coffee, tea, or spices. Imagine: As far back as the 1820s, Bates ran a temperance ship; he wouldn't allow his crew to drink on a voyage across the ocean."

Bates was much older than the other early church leaders, yet he was the only one who stayed healthy, Giang continued. "There's actually a picture from the 1860s of all these church leaders in their twenties who were sick in hospitals and sanitaria, and the guy running the show is this sea captain in his sixties, which back then was really old."

James White had a series of strokes in his 30s and 40s. Ellen G. White took him to a sanatorium, or "water-cure place," and helped him rehabilitate through physical therapy baths and massages and walking. Then she had this prophetic vision that the body is important. "Adventists believe in the body and soul as one," Giang said, "that when you die there isn't a conscious soul floating around, but that

Ellen G. White was one of the founders of the Seventh-day Adventist Church.

you are lying in an unconscious state and God will resurrect you later upon the return of Christ. We truly believe that the body is the temple of the Holy Spirit, and that God communicates to us through our bodies. So the things you do to impair your thinking and impair your health are cutting you off from God's revelation."

Many of the ideas Bates adopted were thus based on health reasons as well as religious ones, and a number of them became centerpieces for Adventist practices, Giang explained. Many focused on preventative medicine. "What strengthened this health message, of course, was that it seemed to work. God wants us to be healthy, and if we did these things, people were healthier."

The Adventists' affinity for health and longevity didn't stop at personal diet and exercise. A respectful emphasis on medical science also became a part of their religious outlook, in part because of their low tolerance for frivolity and their realization that it provided opportunities to evangelize their faith. As much as Ellen G. White and other church leaders embraced the healing done at "water-cure" spas and sanitaria, they frowned upon the dancing and card playing that frequently took place at such facilities and decided to provide a more straitlaced, Adventist alternative.

In 1866, a hydrotherapy clinic known as the Western Health Reform Institute was founded in Battle Creek, Michigan. After doing sluggish business during its early years of

WATER NEEDS

About 60 percent of your body's total weight is water. According to the Mayo Clinic, it takes an average of 8 cups of water (along with a healthy diet) to replace what your body uses normally every day. Moderate exercise increases the amount by 1 to 2 cups. Strenuous exercise (lasting an hour or more) ups the average intake by 2 to 3 cups per hour exercised.

operation, White determined that the place needed a credible medical professional, and chose the smartest young prodigy from a staunch Adventist family she could find, sending him off to medical school at the church's expense. The young man's name was John Harvey Kellogg.

When Kellogg returned to Battle Creek in 1876, he shifted the focus of the institute from water cures to a mixture of Adventist-oriented preventative medicine based on a combination of diet and exercise and cutting-edge medical and surgical procedures. He also changed the name of the place to Battle Creek Sanitarium (a clever variant on sanatorium, the common word for water-cure facilities at the time), which, as business boomed over the next few decades, became known as the San.

Kellogg was a tireless, somewhat eccentric innovator who authored nearly 50 books and was eventually credited with inventing everything from granola and corn flakes to electric blankets and a mechanical horse for exercise—and perhaps peanut butter, which he certainly refined and popularized if not outright originated. By 1888 the rich and famous were making the pilgrimage to the San, which had expanded its dorms and treatment facilities to accommodate more than 600 patients at a time.

Kellogg also had started sanitariums in other places, which perpetuated a demand for more trained people, Giang said. "But again, he and Ellen G. White realized that the worldly medical schools might not offer the same mix of preventative Adventist medicine with traditional therapeutic practices, so he started the American Medical Missionary College in Chicago in 1895." By the end of the century, the Adventists had their own medical and nursing schools. They had also discovered that medical missions could be very effective at reaching people they might not otherwise reach, who weren't necessarily interested in the theology. And it was consistent with the church message of compassion.

Soon, when they went on missions to China or to other countries, they sent medical professionals to perform curative and restorative medicine as well as preachers to teach theology and preventative medicine. "The term Ellen White used was 'the right arm of the church.' That was her phrase for the health message," Giang said. "I think what she meant was that this right arm could reach across cultures and shake hands and open doors. And it also became a good way for paying the bills."

Later Kellogg grew skeptical of White's visions and teachings and was eventually excommunicated from the Adventist Church. He continued to operate the San, which in the early 20th century counted some celebrities among its clientele—including Amelia Earhart, Sarah Bernhardt, Thomas Edison, and President William Howard Taft. While experimenting with shredded wheat cereal, Kellogg and his brother, W. K. Kellogg, accidentally discovered the process for creating flaked cereal. W. K. formed the Battle Creek Toasted Corn Flake Company in 1906. Kellogg was also convinced by his younger brother Will to market his new breakfast foods in what became the wildly successful Kellogg Company, still one of the largest employers in Battle Creek. He died in 1943 at the age of 91.

Ellen G. White, meanwhile, decided to close the Adventist-sponsored medical school in Battle Creek and open another one, to be named the Loma Linda College of Medical Evangelists. CME eventually changed its name to Loma Linda University.

A LITTLE EXERCISE TO STAY YOUNG

Marge Jetton always wanted to be a nurse. "I was working in a cannery when the preacher came and saw me," she said. "He told me, 'Margie, you know you want to be a missionary, and I'll take care of it for you.' Well, that changed my life. I had dreamed of becoming a nurse since I was a little girl, and this was my chance. I feel God arranged for me to get my way."

We sat in Marge's cozy apartment at the Linda Valley Villa, and she kept her end of a bargain we'd made: If I helped on her volunteer rounds about town, she'd tell me her life story.

Marge had been in high gear since 4:30 in the morning. After getting dressed, she did what she called "my devotions," reading lyrics in the song hymnal, and then the Bible. "Every morning," she said, "because who is my friend?" She smiled, then answered her own question. "If you don't have Him, you're out of luck."

After prayers and reflection, it was power-walking time. Marge's corner apartment was in the independent-living facility. If she made her customary beeline down the hallway to the other end of the building and back six times, it worked out to a mile. "I circle through the dining room and get a glass of water each time; six glasses of water before breakfast. Then I come back and get my face on, make the

Marge Jetton, 100, begins every day with a mile walk, a stationary bicycle ride, and some weight lifting. "I'm for anything that has to do with health," says Jetton.

bed, clean up the bedroom, and go to breakfast," she said. Most mornings, she eats oatmeal, but since the Adventists operate Linda Valley Villa, the cafeteria has no shortage of healthy options.

"I can have whatever I want," said Marge. "There isn't anything you can eat that can't be made out of something healthy. My daughter just sent me some waffles made out of soy and garbanzo beans. Mostly I just eat the oatmeal in the morning, and then make up a nice raw fruit and vegetable salad for later in the day."

After breakfast, Marge came back to her room to reflect on the Lord and His blessings. Then she did her exercises. "Today I knew you were coming so I saved some for you," she said. "Let's go."

Marge hit the carpet with both feet before the words left her mouth and led me out the door and down the hallway (past the big, framed poster celebrating her 100th birthday, adorned with dozens of signatures and well wishes) and into the elevator to the basement.

"This is our laundry and exercise area," she said, shooting her left arm out in the direction of the washers and dryers as she wended her way through rows of chairs made festive by brightly colored, patterned ties hanging from the armrest of each chair—props for sitting isometric exercises.

She hopped on the lone exercise bicycle, quickly adjusted the resistance with a spin of the thumb and forefinger, and started churning.

"I ride between six and eight miles a day, except for the Sabbath," she said over the whirring bike. "But yesterday I only rode five because another girl was in a hurry and wanted to get on, so I'll go a little longer today. I set the timer for 15 minutes and try and keep the speed between 25 and 30 miles an hour." She glanced down, saw the speedometer hovering between 20 and 25, and picked up her pace. Then she looked over at me expectantly and nodded at my tape recorder. She began to recount her life story almost nonchalantly, without gasping, never letting up on the pedals.

She grew up poor in what was then rural California, close enough to San Francisco to have a vague memory of water sloshing around in the family animal trough during the great 1906 earthquake. Her father was a mule skinner. Her mother converted to Adventism when Marge was 9 months old, so Marge rarely ate meat and hadn't now for more than 50 years. "My father was a fun man; he'd laugh and pull jokes on us and torment my mother. I'm like him a little bit. I always wanted to do things for people. When I was a little girl I used to tell my mother to go to bed so I could bring her things. I waited on tables when I was 14, and then I worked in a fruit cannery to help the family, but I knew I had to stay in school to become a nurse. My father died of pneumonia before I graduated, but I knew he was proud of me."

Marge met her husband while she was at nursing school in Napa Valley. "He was 16 and I was 18 and he followed me around for three years, if you can call that courting. I told him, 'I'm not going to marry anyone until I'm an RN.' But after that happened, I told him, 'If you really want to be a doctor, we'll get married and I'll support you.'" And she did, through two years of college and then four more of medical school. When her husband was called to serve in World War II for nearly five years, she stayed at home with two small children. ("My children are both 72 now—a son and a daughter we adopted when she was three. I had a hard time having one child and wanted another," she said.)

For 30 years, they lived in Bellflower, California, where her husband was a surgeon and she was a nurse. In accordance with Adventist philosophy, they volunteered together for two international missions, spending three months each in Ethiopia and Zambia.

Always ready to tackle new things, Marge began riding a bicycle at age 12 on a dare from her father, who called her too fat, and she's never stopped pedaling. She took up golf at age 40 and played for decades. At 90, she decided to go to Disneyland and sit in "one of those rides that swirl you around." When she was 96, she finally decided she was tired of cooking and cleaning and moved with her husband into Linda Valley Villa. One day in August of 2003, she returned to their apartment after carrying his food tray back to the cafeteria and he was lying on the floor. "He said, 'I hit my head,' and then his arms went . . .[Marge slumped her shoulders], and he was gone." It was two days before their 77th anniversary.

> ## NUTS FOR NUTS
>
> Looking for a way to get a daily serving of nuts? Add toasted whole walnuts or pecans to a green salad. Try roasted cashews in a chicken salad. Finely chopped nuts can also make a delicious coating for fish filets, but popping them right in your mouth works, too.

CONVERTING STRANGERS TO FRIENDS

"It took me a year to realize that the world wasn't going to come to me," Marge said. "That's when I started volunteering again, and it was the best thing to ever happen to me. I found that when you are depressed, that's when you do something for somebody else. There, that's eight miles," she said, abruptly changing her mood and the subject. "See what you can do on this bike—one of the girls here taught me this." She cackled, suddenly pedaling furiously in the other direction. "It uses a whole different set of muscles."

"I can't believe you aren't the slightest bit winded," I told her. "I would be."

"Really?" she answered, and in the space of a few seconds I saw three reactions move across her face: contempt that I'd let my body go, skepticism that I was pulling her leg, and then genuine concern that I was telling her the truth. "Well then, let's get you going," she announced, hopping off the bike and moving toward a series of hand weights in the corner. "Pick up a pair of five-pounders; that's what I'm using. Now go sit in that chair." She demonstrated an eight-part, dual-hand, lift-and-stretch exercise as she counted it off, and barked for me follow along. Then we started with twisting trunk contortions, followed by side-to-side stretches with the weights held over our heads. "I've got to go to the bathroom," she said, after about ten brisk minutes. "Why don't you take a little ride on that bike while I'm gone?" Although I can't be sure, I think I heard her linger an extra minute or two out of sight in the basement hall while I dutifully pedaled.

"All that water," she said by way of explanation as she returned. "C'mon, I want to show you our garden." On the way up, I asked if she ever got lonely. "Well, sure, you miss people. Most of my friends have died. My husband is dead," she said matter-of-factly. "But I just like to talk to people. My motto is: A stranger is a friend we haven't met yet. I don't know a lot of non-Adventists. I don't go to movies or dances. But I still have a lot of friends around this place and other places in town. I think I was probably shyer when I was younger. There are days when my forgetter is working overtime. I remember just as many things as I did before, but after a hundred years or more, I have so many things in my computer I don't know what button to push." She walked off the elevator laughing at herself.

"Look at this: I have an ear of corn! This is the first time I have ever grown corn," she said, as we walked in the courtyard. Plants, flowers, and vegetables were growing in long rows against all three walls of the fence that squared off the yard from the main building. "These are my tomatoes—oh, look, I've got a ripe one! See that hydrangea over there? When I first came here, there were some as tall as me. Then somebody cut 'em down, and now they are growing back."

Some people walked by and Marge waited until they were out of earshot, then said, "There are some folks in here who are 80 years old. I was out still working when I was 80. I tell them, 'I'm old enough to be your mother!'"

"Ahh, the roses," she gushed, leaning forward into a brilliant red trio of them. "I can't look at a rose without thinking of Jesus. When I was a girl, we used to go on long family walks on the Sabbath and look at all the wildflowers. It was such a special time. And the Sabbath still is special you know," she said, finally stepping back from the flower. "It is a time to come apart from the things of the world. It is something to look forward to. You get to go on hikes, stop pushing. I don't know how to describe it to you. I think the Sabbath gives you peace, and that contributes to your health."

A few minutes later, it was time for me to go. When I complimented Marge on her firm handshake bidding me adieu, she replied, "I'm a strong woman. That comes from all the massages I have given in my life. Say, you need to remember to send me what you write. People come and talk to me because I'm old, and that's the last I hear from them. They probably think I'm going to die, or have already died. But I'm still here."

SANCTUARY IN TIME

Inspired by Marge's pep talk about the Sabbath, I decided to attend an Adventist service at the church on the Loma Linda University campus. Adventists observe the Sabbath from sundown on Friday to sundown on Saturday. The number of Adventists in Loma Linda is high enough to warrant that postal carriers deliver mail only from Sunday through Friday.

At first, the tenor and content of the service itself didn't seem so different. As in many "mega-churches," video cameras swung over the pews on boom cranes, their images shown on large screens over the stage as well as being broadcast to viewers at home. Hymns were sung. A sermon was delivered. The collection plate was passed. I wasn't sure what I was looking for—signs of why this faith promotes a longevity culture, I guess—but I wasn't finding it.

After the service, the hundreds in attendance seemed to hang around longer than usual, and in larger groups. It was the social equivalent of comfort food. One cluster of eight to ten young people, a diverse mix of race and gender but all between their teens and their thirties, had locked arms and were standing around the Good Samaritan sculpture near a side entrance of the church. More than half of them were chanting the biblical parable cited on the plaque of the sculpture, from Luke 10: 25-37. The passage is about helping the downtrodden.

But what was striking was not the message, so much as the variety of expressions. One student was reading from a Bible; another had her eyes shut and was reciting it from memory; two others were only occasionally chiming in, and mostly doing can-can kicks in clownish, spontaneous cho-reography. It was an ad hoc, perhaps fleeting community, and everyone seemed happy. It reminded me of what Marge Jetton had told me about why she volunteered so much: "Because it makes me feel good," she'd said. "Don't you like to help people up when they need a hand?"

Randy Roberts, the pastor of Loma Linda University Church, told me more about the spirit of the Sabbath and why he thinks it helps people live longer. "I think even for those who may slip away from the faith and from the Adventist Church, observing the Sabbath remains a linger-ing reality, a way to stay connected," he said. "It is meant to be a sanctuary in time for rest and rejuvenation, and I think it accomplishes that on a number of levels.

"Ellen White had this whole-health emphasis that you should use the time to value exercise, get out in nature, and move around. Another way it is healthy—I don't have any studies or hard numbers for this, just a lifetime of observing it myself—is as a pure stress reliever that allows some peace to occur."

"I've heard over and over again from students in rigorous programs like medicine and dentistry, and from faculty too, that they can't wait for the Sabbath to come because they have a guilt-free time when they don't have to study or do some other obligation. They can just be with their family and friends and with God, and just relax and rejuvenate. When you have that as a pattern in your life 52 times a year, it can make a big difference. Some call it a 'sanctuary in time.' Another part of Adventist beliefs is that the Sabbath reminds us we are creatures and not creators," he said. "It reminds us that we don't need to have all the answers, that we recognize our finite capabilities, and that we are dependent on God. That also is part of the sanctuary."

Roberts said he'd seen studies showing that people with two or three significant ties in their lives, to family and friends and community, tended to be healthier, both emotionally and physically. "The Sabbath gives most Adventists a time to do that: to shut off the television, not think about your work or business, and just spend time with the people who are important to you. Unfortunately, this is becoming more unique. You used to see it more in other faiths that set aside Sunday—it would be like old *Andy Griffith Show* reruns. That doesn't happen as often now. But it is still an important part of our faith."

OPEN-HEART SURGERY ON TUESDAYS

It was a hot, breezy afternoon on a rare, smog-free day in the San Bernardino Valley when I dropped in on Dr. Ellsworth Wareham, who was hard at work in his backyard. From his property on Crestview Drive you can look out on the valley's mountains receding like brown waves, and when the air is as clear as it was that day, you could even make out the shirt colors of people walking on the LLUMC campus a mile and a half away.

For all the sweat rolling into his eyes, however, Wareham couldn't see much of anything. When he wiped his brow with his forearm, it was as if he had used a squeegee to send water down the rest of his body, soaking his clothes. His shoulder muscles showed through his clinging T-shirt as he struggled to corkscrew the post-hole digger through the packed soil. Recently, a contractor had presented Wareham with an estimate of $5,000 to install an eight-foot-high wooden fence along a steep hillside on the edge of his property.

By consulting his local discount hardware store, he had learned that the fence materials would cost only about $2,000. So he decided to do the work himself. Thus far he'd sunk a couple of poles and sealed a couple of holes, but there was an imposing pile of posts and fence sections left. There was also a potentially significant complication: Ellsworth Wareham was in his 90s. He was born in 1914.

Four days later, Wareham was in open-heart surgery at a community hospital on the edge of Los Angeles. But he was not the one on the table. He had a scalpel in his hand.

"I am a man of good fortune," he said, when we next met for our first extensive conversation. "My hands are still steady, my vision is good, and I don't have Alzheimer's disease, which may occur in about 50 percent of the people my age.

"Now, I want to make it clear that I am no longer the lead surgeon; I am either the first or second assistant, and I do the minor parts of the operation. If I had to, I could step in and do the major parts—that's one of the reasons I'm in the room. But cardiac surgery is inclined to be long and sometimes quite taxing, lasting anywhere from three to six hours.

"Sometimes I'll go out and rest for a half hour, and that's very refreshing. To increase my endurance I exercise regularly by doing my own landscape maintenance and gardening. But in recent years, in spite of adequate exercise, my

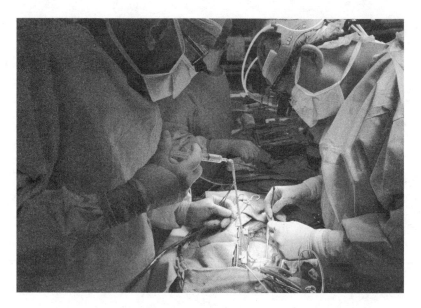

Dr. Ellsworth Wareham, 91, assists during a heart surgery procedure,
something he does about two to three times per week.

stamina is decreasing. The hormones that build the muscles
are diminished.

"So yes, I am sensitive about my age," he said. "If someone tells you the assistant on your heart surgery is in his
90s, well, that's not a comfortable thought. But the way
I see it, the cardiac surgeons who want me there to assist
them can judge my ability better than I can. And when I
start to do less than top-quality work, I trust them not to
use me." Eventually, he won't renew his license, he added.
"And a deficit in mental or physical ability may appear at
any time and make the decision for me."

As I parked on the steep hill near his driveway, I spotted
Wareham on his hands and knees, cleaning leaves out of
the gutter in the road. Tall, thin, clad in a short-sleeve red

shirt and khakis, he greeted me with a firm handshake and a ready smile. His face was framed by a mustache and hair that was balding in the middle but thick on the sides. As we entered the house to the cacophony of three dogs (a mutt, a golden retriever, and a Chihuahua), Wareham's wife was unpacking suitcases from a trip to Canada to celebrate the centennial of tiny Canadian Junior College (now Canadian University College) in Lacombe, Alberta, where he graduated from a two-year premedical course in 1933. Wareham had driven them home from the airport last night.

"I think it is important for me to keep active," he said, gesturing for me to take a seat in his study. "People say, 'Oh, I don't drive at night.' Well, I drive over 2,000 miles a month on Southern California freeways, much of it at night. I think it keeps me alert." Through the double doors over his shoulder and past the patio, I noticed the long, regal, wooden fence in the backyard.

The walls of Wareham's study are laden with prestigious academic and honorary degrees, along with photographs from his missions abroad. As a doctor in the Navy during World War II, he once took out an officer's appendix on a destroyer being tossed around in a typhoon off the coast of Okinawa. Later, on a medical mission to Pakistan sponsored by the U.S. State Department in 1963, he was a surgeon on the Loma Linda heart surgery team that brought open-heart surgery to that country for the first time. He pointed to a photograph of a girl who, with her family, had walked 100 miles to get an operation. In Vietnam, a year before South Vietnam fell, the work of the heart surgery team in Saigon was featured on the Walter Cronkite TV program.

As a surgeon he was a pioneer of open-heart procedures. (Dr. Leonard Bailey, who has performed more infant heart transplants than anyone in the world, unabashedly refers to Wareham as "my mentor.") When rudimentary open-heart surgery was first practiced in the early 1950s, Wareham saw the emergence of a new field and extended his residency to acquire the necessary expertise. But what he saw on the operating table would have a profound effect on his own health habits.

"In the early days, when we used the heart-lung machine, we connected the arterial line to a cannula in the leg artery, later it would be straight into the aorta," he said. "But I observed when I was cutting into the thighs of these patients that those who were vegetarians had better arteries."

"When we did the surgery," he continued, "if it was a nice, smooth artery, I went back later and asked the patient, and it turned out that he or she was a vegetarian. And those who really had a lot of heavy calcium and plaque in the arteries, their diet would not be toward the vegetarian side. Now that wasn't true 100 percent, and I didn't keep any statistics or write any papers or anything; it was just something I observed. But I began thinking about it. And I saw people getting their toes cut off or their feet cut off because of vascular disease, and that motivated me. So it was a gradual thing." In middle age, he decided to become a vegan.

"Now, I must admit, it is no problem for me to be a vegan," Wareham said, leaning forward. "My mother was an Adventist, and my father was not. I grew up on a farm, and I milked the cows, and I never did care for milk. I don't particularly care for eggs either."

"For quite a few years, I read the concerns about vegans not getting enough vitamin B$_{12}$ or protein or calcium; you know how it goes," he said. "They said even the amino acids in vegetables were not adequate. But we began to find out that much of these were old wives' tales. With the exception of lack of B$_{12}$ being of some concern, you aren't going to become deficient in protein and all these things. I use soymilk, and as far as eggs are concerned, my wife knows suitable substitutes.

"My wife is not a vegetarian, but she is changing. Again, it is no credit to me that I am a vegan—I just love it. There are so many tasty fruits, vegetables, and nuts. This morning we had wonderful strawberries, for example."

The longer Wareham talks, the more apparent it became that he's a walking advertisement for the Blue Zone lifestyle of the Adventists. "I am very fond of nuts, all kinds of nuts," he said happily. "I have to restrain myself. Most days I eat two meals, first around ten in the morning and then again around four in the afternoon, so I can keep my weight down. When I eat, I really enjoy it, and twice a day is enough. Nuts are usually part of the menu. I know walnuts are reported to be very good for you, but I don't eat them because I enjoy peanuts and cashews and almonds so much. Sometimes I get purist and think I should eat them

EXERCISE TYPES

A combination of four types of exercise will keep the body balanced and strong. Endurance: Activities like walking, hiking, swimming, and cycling improve the health of the cardiovascular system. Strength: Lifting weights builds up and maintains muscles. Flexibility: Stretching keeps us limber and flexible. Balance: Practicing balance through activities like yoga will help avoid falls.

raw, but really, whatever is handy. I am not a nutritionist, and I don't profess to be."

Another thing that helps keep the weight off is drinking water, Wareham noted, as, almost in gentle rebuttal, his wife entered and gave us both a glass of cranberry juice. "I became aware some years ago that water is highly important to health, and I do make an effort to drink a lot of it. I'll drink maybe three glasses of water when I first get up, because I want to make sure before I get busy and forget to have some. Then when I get home I have some more. And one of my little rituals is to never pass a water fountain without having a drink. It adds up."

"But there is something else," Wareham said, lowering his juice glass just before he was about to take a sip. "I know people that have had problems who are just as careful in health habits as I am. There's a colleague my age, who in addition to his M.D. has a doctoral degree in public health from Harvard University, and he has followed a good program—he walks, eats right. Now this fellow has had cancer of the prostate and the neck, and two heart attacks, each followed by heart surgery and bypass grafts. Why wasn't he protected? Luck? Genes? Maybe stress. Remarkably, ten years after the last onslaught of these diseases, he is alive and active. I would conjecture his good health habits helped him recover. I know that it really helps me to have purpose to what I'm doing, to have a job and finish the job and have another job, and to enjoy that process. As I say, I am blessed." But the tone was not totally content this time. It was hard for me to tell whether Wareham was gloating, feeling guilty, or experiencing a little of both.

"A few individuals relating anecdotes about their life-styles and health histories do not prove anything. But the Adventist health study by Dr. Fraser and his coworkers is a highly significant work," he said. "It examines lifestyles and health outcomes of more than 34,000 people over a period of 12 years. The study should have serious consideration."

As we negotiated our way through the dogs and walked toward my car, Wareham closed with his own little pep talk. "It is so simple just to get on a good program. I don't have great genes. I just try to observe broad principles that any-body can pick up, and I still don't take any blood pressure medication. We are especially blessed out here in Califor-nia, where there are so many good things to eat."

"People talk about curing cancer and heart disease, and of course it is an important and worthy goal that can't happen soon enough. But there are simple things everyone could be doing right now that would save so much money and suffering—like drinking enough water every day, exercise, and eating healthy food. But hey," Wareham said, suddenly catching himself in his fervor, "everybody has his own idea about these things—it's their lives, after all. You can tell somebody what to do, but it's up to them whether they do it. But you can tell them how good *you* feel."

PART OF THE CLUB

Shortly after three on Friday afternoon, I visited the Mock family in Yucaipa, a burgeoning hillside town about seven miles east of Loma Linda. I was hoping to see how a typi-cal Adventist family observes the Sabbath to witness how it might contribute to a culture of longevity. When I knocked on the door of a large new suburban home on a cul-de-sac, the door opened and I was met by three males—Jesse Mock and his two sons, Justin, 15, and Austin, 13—who looked remarkably alike, with lithe, athletic builds, full faces with toothy smiles, and thick, fair, curly hair. Jesse's wife, Rhonda, greeted me from the kitchen, where she was getting dinner ready.

Inside, the conversation flowed easily, as Jesse and Rhonda told me how they had just moved here a few months ago from Denver, Colorado, so Jesse could take a job as a vice president at LLUMC. They'd lived in the area years ago, and were happy to be back in an area with a robust church school program, which gave them options after a positive, six-year experience with home-schooling. (Adventists believe in a parochial education and operate a large, church-sponsored school system in the United States.)

The family was vegetarian, Jesse informed me, "But I think the boys have had meat before. Remember that birth-day party, Justin?"

"Yeah, I've had meat," Justin said with a shy smile, and added, with a tone of surprise, "It wasn't that bad." Austin disagreed. "Uhhhh, I didn't like it much. I don't know. It just didn't taste right."

The doorbell rang, and it turned out to be Dr. Daniel Giang, with whom I talked earlier at LLUMC, and his wife and two children. His son and the two Mock boys dashed outside to play basketball, while his younger daughter stayed behind with the adults. Again, the conversation flowed eas-ily. The women caught up on the past week, while Jesse

A shared family meal often begins the Seventh-day Adventists' Sabbath.

and I took turns explaining how we knew Dr. Giang. Ever since they moved to Yucaipa, the Mocks have received helpful advice from the Giangs about local schools, recreational opportunities, and all the other aspects of daily living in the community. That both Giangs are medical doctors and both Mocks are not seemed less important than their common religion and the similar ages of their sons. After about 15 minutes, Jesse called in the boys, and the families sat around the living room to listen to a brief Bible passage. Then it was time to eat.

Everyone grabbed a plastic plate and loaded up in the kitchen before heading to the dinner table. There were soy "chicken" salad sandwiches in small, not quite bite-sized buns, and a beautiful salad that included some of the largest, most succulent blackberries I've ever seen. I enjoyed the sandwiches much more than I expected, and was a little embarrassed to realize that some of it was the presentation: The soy was served American style, slathered on a bun with some celery for crunch, and perhaps some mayo for creamy richness. Among the desserts was dried hibiscus, something Rhonda picked up on a lark, she said. It tasted candied, like dried apricots or mangoes.

The Giangs lingered a while after dinner, and the two couples gossiped about the university and the Giangs' daughter's dance classes. There was also some intelligence sharing about which sports teams the sons should try to play for in summer.

After the Giangs left, I was invited to accompany Justin and Austin to a rehearsal, or jam session, with other youth members of the church the Mocks attend, as they prepared for a performance at the service the next morning. Austin's drum set was loaded into the van. Jesse drove over with me, talking mostly about college basketball and how the boys have adjusted to the move.

The ten or so members of the youth band aren't about to top the charts anytime soon. But it was entertaining listening to the music director good-naturedly cajole slight refinements on two or three takes of a song, while his charges alternately ignored or obeyed him with about equal results—especially with the parents gathered around the back of the garagelike rehearsal room. Two coltish girls, close friends and tandem vocalists, spiced things up by acting out some of the lyrics. For Bill Withers's "Lean On Me," for example, they leaned against each other on the chorus and flexed their biceps on the line, "when you're not strong," and cradled their arms for imaginary babies during, "I'll help you carry on."

So this was a slice of the Adventist Sabbath. If you're not part of the community, it might seem almost quaint. But for members of the community, this kind of repetitious, casual intimacy provokes good memories and a subtle sense of safety and well-being—the same kind of constant, subtle reinforcement of healthy habits that seems to give Adventists a better chance for longer life. It's no wonder that most Adventists seem to hang out with other Adventists.

HOLD THE BACON

I'm never going to be an Adventist. But you don't have to convert to their faith to admire how they have generated a Blue Zone out of whole cloth by sticking together and reinforcing the right behaviors for longevity. Having spent more than a week in Loma Linda, I'd met some amazing characters: Marge Jetton and Ellsworth Wareham, of course, and more.

Just down the hill from Ellsworth's place was a rambler owned by Minnie Wood, who will turn 100 in May 2008. Minnie was a 1925 beauty contest winner and said she would have gone into acting had she not converted to Adventism. Yet she had no regrets. "How many actors have gotten to perform in front of three presidents?" she thundered, ticking off Truman, Eisenhower, and Nixon as those who heard the choirs she directed while living in Washington, D.C. Minnie taught music while traveling with her husband in China and Singapore. She continued to teach some students, driving to their homes in her Oldsmobile, and could still carry a tune as long as she could synchronize the key on the piano. "I love what I do. I always have. I wouldn't do anything else."

There was Letha Graham, born in 1906, who lived in a home at the top of a mountain ridge and rented out rooms to patients undergoing proton radiation treatments at the medical center. Letha used to sell medical encyclopedias on horseback in Alberta, Canada. In recent times she lived with her 94-year-old sister, taking a daily walk around the house grounds, and pausing for a dip in the Jacuzzi.

LOVE TO LAUGH

Studies have found that a belly laugh a day may keep the doctor away. In 2005, researchers at the University of Maryland showed that laughter helped relax blood vessels, linking it to healthier function and a possible decreased risk of heart attack. Others have found that laughter may lower blood pressure and increase the amount of disease-fighting cells found in the body.

Just down the hall from Marge Jetton at Linda Valley Villa was 94-year-old Dr. Wayne McFarland, creator of the first—now famous and highly successful—five-day anti-smoking plan. A wiry firecracker of a man, McFarland slapped things— his hands together, the back of a chair, your shoulder—for emphasis when he was excited. Remembering when he and Dunbar Smith were both staff physicians at the Battle Creek Sanitarium and Hospital conducting five-day plans to stop smoking, Wayne recalled finding two pamphlets, "More Nuts & Less Meat" and "The Simple Life in a Nutshell" by Dr. Kellogg. Indeed, the apartment he shared with his wife had numerous bowls filled with nuts. Later, McFarland took me out to the same garden where Marge showed off her corn and tomatoes, and he slapped the cement wall with glee when he noticed how much his own tomatoes had ripened.

Then there was Ethel Meilicke, 108, who lived in the nursing home part of Linda Valley Villa. With a history remarkably similar to that of Marge Jetton, she grew up on farms, worked in a cannery, and became a nurse as a young woman. When I visited, she was frail and hard of hearing, but when I asked what she enjoyed most during her days on the dairy farm, she blurted out the word "shneckens," a German pastry her daughter (now 80) said she always used to make. Ethel passed away less than a month after my last visit to Loma Linda, on July 1, 2007.

It still amazes me that Ethel Meilicke, Wayne McFarland, Minnie Wood, Letha Graham, Ellsworth Wareham, and Marge Jetton all lived within a three-mile radius of one another, the youngest of them 93 (and still practicing open-heart surgery). As I prepared to leave Loma Linda, I remembered something Dr. Fraser from the AHS had told me. "Some Adventists get personally offended if they get colon cancer or some other disease," he said. "They have a reputation for avoiding these things now, of course, but it begs the question, what do you expect to die of? And when we looked, we found that, by and large, the proportions of deaths from different causes in Adventists are about the same as everybody else. It is just that they die later."

What does that mean? Are Adventists hitting some kind of fundamental rate of aging? "We see people get to age 98, or 103, or whatever, and as physicians we ask, what did they die of?" Fraser said. "We see they were experiencing heart failure, and their kidneys weren't so good, and everything was slowly deteriorating. So aging seems to be about deteriorating cellular function and metabolism and so forth. And what I am saying is that with the Adventists it almost looks as if that general deterioration and cellular function may be impacted by lifestyle. If so, that would be very interesting."

My last stop in Southern California was a pancake house. Tempting pictures of maple syrup slathered over sausage and pancakes adorned the menu. I was hungry, and the waitress had her pencil poised. "I'll take the oatmeal with fruit," I said. "And just water." We'll see how long that lasts.

LOMA LINDA'S BLUE ZONE SECRETS

Try these tactics practiced by America's longevity all-stars.

Find a sanctuary in time.
A weekly break from the rigors of daily life, the 24-hour Sabbath provides a time to focus on family, God, camaraderie, and nature. Adventists claim this relieves their stress, strengthens social networks, and provides consistent exercise.

Maintain a healthy body mass index (BMI).
Adventists with healthy BMIs (meaning they have an appropriate weight for their heights) who keep active and eat meat sparingly, if at all, have lower blood pressure, lower blood cholesterol, and less cardiovascular disease than heavier Americans with higher BMIs.

Get regular, moderate exercise.
The Adventist Health Survey (AHS) shows that you don't need to be a marathoner to maximize your life expectancy. Getting regular, low-intensity exercise like daily walks appears to help reduce your chances of having heart disease and certain cancers.

Spend time with like-minded friends.
Adventists tend to spend lots of time with other Adventists. They find well-being by sharing values and supporting each other's habits.

Snack on nuts.
Adventists who consume nuts at least five times a week have about half the risk of heart disease and live about two years longer than those who don't. At least four major studies have confirmed that eating nuts has an impact on health and life expectancy.

Give something back.
Like many faiths, the Seventh-day Adventist Church encourages and provides opportunities for its members to volunteer. People like centenarian Marge Jetton stay active, find sense of purpose, and stave off depression by focusing on helping others.

Eat meat in moderation.
Many Adventists follow a vegetarian diet. The AHS shows that consuming fruits and vegetables and whole grains seems to be protective against a wide variety of cancers. For those who prefer to eat some meat, Adventists recommend small portions served as a side dish rather than as the main meal.

Eat an early, light dinner.
"Eat breakfast like a king, lunch like a prince, and dinner like a pauper," American nutritionist Adelle Davis is said to have recommended—an attitude also reflected in Adventist practices. A light dinner early in the evening avoids flooding the body with calories during the inactive parts of the day. It seems to promote better sleep and a lower BMI.

Put more plants in your diet.
Nonsmoking Adventists who ate 2 or more servings of fruit per day had about 70 percent fewer lung cancers than nonsmokers who ate fruit only once or twice a week. Adventists who ate legumes such as peas and beans 3 times a week had a 30 to 40 percent reduction in colon cancer. Adventist women who consumed tomatoes at least 3 or 4 times a week reduced their chance of getting ovarian cancer by 70 percent over those who ate tomatoes less often. Eating a lot of tomatoes also seemed to have an effect on reducing prostate cancer for men.

Drink plenty of water.
The AHS suggests that men who drank 5 or 6 daily glasses of water had a substantial reduction in the risk of a fatal heart attack—60 to 70 percent—compared to those who drank considerably less.

Discovering Costa Rica's Blue Zone

Discovering Costa Rica's Blue Zone
Tortillas and Beans, Hard Work, and Something in the Water?

I T WAS SEVEN IN THE MORNING IN THE village of Hojancha, Costa Rica, and already the tropical sun was bearing down as I opened the squeaky front gate and called into the pastel-pink house for Tomás Castillo—Tommy to his friends.

A moment later, Tommy emerged with his big, white-toothed grin, his brown pants tucked into his boots, a T-shirt, and his New York Yankees baseball cap pulled down low over his forehead. "*Listo?* (Ready?)" he asked, with no intention of waiting for an answer. We mounted a pair of bikes and, with a truant's delight, rolled out into the hot Costa Rican morning.

Tommy leaned forward over his handlebars as we swooped down the street with the wind whistling past our ears and the blurred line of aqua, pink, and ocean-blue houses swooshing by. We squeezed our hand brakes and slowed at the town square, where men at the Café Central eating their *pico de gallo* and eggs called out to Tommy. He waved back, blushing. I wasn't sure if he was proud or embarrassed to be cycling with a six-foot-tall gringo. On the plaza corner, a pushcart vendor squeezed fresh orange juice and sliced mango and papaya for the morning's customers. We turned right.

Our route took us by the town clinic, past a mechanic where the rhythm of local cowboy music blared into the street from tinny speakers. We swooped down another hill past the village school, and from there, the houses thinned out. On one side of road, buildings gave way to a wall of jungle. The road dipped to where the pavement bridged a creek and continued up a steep incline. Tommy stood up hard on his pedals and pulled ahead of me. I was breathing heavily. Sweat trickled down my back.

We followed an unmarked dirt road into the forest. Our bike wheels traced parallel ruts past a horse barn and a vegetable garden. The road stopped at a clearing with a raised chicken coop, a tin-roofed wooden house, and a woodshed stacked high with split logs. Out front, a woman wearing a bright pink dress, hoop earrings, and carnival beads vigorously swept the jungle floor, sending up a dust cloud. Behind her, a few long, golden pencils of light angled through the trees.

"*Hola,* Mamá!" shouted Tommy as he dismounted his bike. Tommy's mother—Panchita—dropped her broom in surprise and gleefully greeted her son with an embrace, and then turned to me. "OyEEE, God blesses me!" she exclaimed. "I have foreign visitors!" Then she hugged me.

She took us both by the hand and led us to her porch, where she jumped up on a bench and dangled her legs in the air. It was only 7:30 a.m., but Panchita was ready for her midmorning break. She'd been up since four and had already knelt next to her bed to say her morning prayers, fetched two eggs from the chicken coop, ground corn by hand, brewed coffee from well water drawn from the limestone bedrock beneath her house, made herself a breakfast of beans, eggs, and tortillas, split wood, and using a machete almost as tall as her five-foot frame, cleared the encroaching bush around her house. She asked if she could prepare breakfast for us. "No," said Tommy, who was sweating lightly under his baseball cap. "I'm not hungry."

"Oh, you know better," Panchita scolded. "Let me make you some eggs." And she jumped off the bench.

"No, no, Mamá," Tommy said, shifting uncomfortably on his bench. "I'm fine."

Panchita pulled herself back up and now began to stroke Tommy's knee. "How is your leg, my son?" A few days earlier, he had injured it working around the house.

"Mamá, I'm fine, please!" he said, grimacing.

As the scene unfolded, I sat by and smiled to see an exchange between a loving mother and a son who didn't want to be embarrassed in front of a new friend.

Under the circumstances, I could see Tommy's point. He was, after all, an 80-year-old man and a great-grandfather himself. His mother, Panchita, had recently celebrated her 100th birthday. Hojancha, where they live, has one of the healthiest, longest-lived populations on the planet—a place where sons can take their time growing up.

SEARCHING IN CENTRAL AMERICA

In 2002, as demographer Luis Rosero-Bixby was doing routine work with Costa Rican population data, he noticed that men here seemed to be living longer than those in more developed countries around the world. This remarkable fact had gone unnoticed because in developing countries and in Central America—a part of the world notorious for malaria, dengue fever, and revolutions—most mortality studies didn't even ask if anyone lived past age 80, which was considered well beyond the life expectancy of the area.

Moreover, organizations like the United Nations had assumed that many Costa Ricans often exaggerated their ages, so that any finding would be considered invalid. Nevertheless, Rosero-Bixby, the director of the Central American Population Center in San José, decided to investigate further. Instead of looking at census data (in which respondents can misreport their age), he used a forgotten but foolproof method of determining how long a population of elders lives. The "quasi-extinct cohort" method was based on using the well-documented death records of Costa Rica's voting population and birth records to compute ages. (Unlike many places in Latin America, the depoliticized office of statistics and census has been free of government meddling since 1883.)

Put simply, he took a sampling of births recorded between 1890 and 1900 and then found the death records. From that, he calculated the average age of death (life expectancy) and the chance of dying at any given age (mortality rate).

Comparing these findings with data from developed countries, Luis figured that a Costa Rican man at age 60 had about twice the chance of reaching age 90 as did a man living in the United States, France, or even Japan. He also found that if a male reached 90, he could expect, on average, another 4.4 years of life—again a life expectancy higher than in most developed countries. If Rosero-Bixby's numbers were correct, it was an extraordinary find. Costa Rica spends only 15 percent of what America does on health care, yet its people appeared to be living longer, seemingly healthier lives than people in any other country on Earth.

In 2005, Rosero-Bixby traveled to France to present his findings to a group of his peers at a world congress on population studies. A subset of the demographers at the congress study long-lived populations, and many attended Rosero-Bixby's talk. They listened attentively, but few seemed to trust his findings: Infectious diseases and political instability are known to shorten people's lives in Central America. How could these demographers be expected to believe that any of the region's populations could outlive their counterparts in more developed countries? Rosero-Bixby received polite applause, and then the audience scattered. He returned to Costa Rica and continued his work, mostly focused on population growth in Central America, and his longevity findings went on the back burner.

Not long after the conference, I phoned demographer Michel Poulain. My article in NATIONAL GEOGRAPHIC, "The Secrets of Long Life," had met with success, and I was curious to find more of the world's undiscovered Blue Zones. I knew that Michel maintained a personal database of the world's oldest populations, and I figured he might have some good ideas.

"Yes, there are other places" he said in his heavy French accent, and he told me about his recent travels to Crete and Majorca to investigate claims of longevity. "But not so special as Sardinia."

I asked him if he'd ever considered places in the developing world. "The problem," he replied, "is that these places have usually very poor record keeping, and it's impossible to verify ages." He paused. "Maybe," he said in the voice that accompanies an "ah-ha" moment, "there is one place."

Michel had attended the world congress on population studies, and he told me about Rosero-Bixby's paper. He said that Rosero-Bixby had struck him as a good scientist whose data had perhaps been dismissed too quickly. I made a deal with Michel. If he'd examine Rosero-Bixby's data and validate it, I'd find the money to finance an investigative trip.

After our initial conversation, Poulain contacted Rosero-Bixby, explained our work in other Blue Zones around the world, and said we were interested in searching for a Blue Zone in Costa Rica. Would he work with us? He agreed. In fact, Michel checked his detailed data and, based on it, identified a group of villages around the Nicoya Peninsula where the proportion of the oldest people was significantly higher than in the rest of the country.

Rosero-Bixby agreed with this, as he had already begun work on comparing different regions in Costa Rica using a technique that involved drawing thousands of circles on a map of Costa Rica and checking the life expectancy and mortality rates of all the people living in those circles. Indeed, it produced one region in the northwest of the country including the Nicoya Peninsula that was significantly different from any other (e.g., in that region, people die of cancer at a rate 23 percent lower than in the rest of the country). Accordingly it was agreed that another potential Blue Zone might be identified and that these preliminary indications justified a trip.

A NEW BLUE ZONE?

Eight months later, with a grant from National Geographic's Expeditions Council, Michel, videographer Tom Adair, and I arrived in San José for what amounted to a week-long blind date. On a steamy Tuesday morning in August, our taxi wove through a labyrinth of streets near the University of San José and came to a stop at an impressive, three-story building on the University of Costa Rica campus.

Inside the Central American Center for Population Studies, Rosero-Bixby was waiting for us. A pleasant-looking man of 54 with graying black hair and wire-rimmed glasses, he was dressed in khakis and a polo shirt. He grabbed our hands and greeted us in perfect English and with the ebullient hospitality characteristic of Latin America. Did we want lunch? Coffee? Was our hotel reservation okay?

He had taken three days' vacation and borrowed the Center's Toyota Land Cruiser to join us on an exploratory journey to Nicoya. He had also personally mined his databases to find the names of 30 people over 90 years old and had their profiles and photographs loaded in the PDA he wore around his neck.

He and Michel had mapped out a plan: We'd travel to Nicoya, interview a random sample of at least 20 people over 90 years old to verify their ages, and get a feeling for their lifestyle. Then we'd return to San José and further verify their ages in the national archive. If ages checked out, we'd assemble a bigger team and return on a second trip to try to explore further why people were living so long.

Our route out of San José took us north along the Pan-American Highway. I remembered the road well. Two decades earlier, I had traveled it from the opposite direction on a bicycle expedition from Alaska to Argentina. Then it had snaked pleasantly through the forested mountains that rose upward from the Nicaraguan plains. Few cars passed and even fewer trucks. Unlike the Pan-American Highway in most other countries, the road here cuts through very few villages (a surveyor had told me back then that a wealthy landowner had convinced government officials to route the highway past his land instead of through a lower, more populated route). Now, long trains of diesel-belching semis clogged the road, heaving slowly up the hills and coasting cautiously down the other side.

But for Luis at the wheel, this was travel as usual. He was telling us the story of his childhood in Ecuador and how he had fallen in love with a Costa Rican who'd brought him to San José, where he'd worked as a lecturer in demography and learned English.

He made his way to the United States, earned a Ph.D. in demography, and then secured a position at Princeton

University where he worked in the office of Office of Population Research. He established the Central American Population Center with grants from the Bill and Melinda Gates Foundation in 2000 and has run it ever since. "It's really nothing," he concluded with a modest shrug.

At Las Juntas, three hours north of San José, we turned off the Pan-American Highway and drove westward through progressively drier and hotter terrain. The forests eventually gave way to cow pastures dotted with humped-back Brahman cattle and occasionally the massively crowned guanacaste trees after which the region is named.

We crossed the Taiwan Friendship Bridge spanning the alligator-infested Tempisque River and onto the roughly 80-mile-long finger of land south of the Nicaraguan border along the Pacific coast—the Nicoya Peninsula. "Until very recently, this was one of the most isolated parts of Costa Rica," Luis said. "You can see it's a long way from the highways, and before this bridge was built, most people had to take a ferry to get here."

This in fact was our first clue to Nicoya's longevity. Sardinia's Blue Zones had been similarly isolated. Could the population here have been isolated and then evolved a genetic makeup that favored longevity as had happened in Sardinia, we wondered? "Probably not," replied Luis. "It's a pretty mixed-blood race living here."

"What else distinguished the people here from everyone else in Costa Rica?" I asked.

Luis thought for a minute. As a demographer, he tended to be focused more on numbers than on the people behind those numbers. He had led teams of surveyors who had

All Costa Rican citizens carry a cedula de identidad, *an identification card listing both name and date of birth (making verifying the centenarians' ages much easier).*

interviewed people, but that was mainly to compile survey data. The idea that he had stumbled upon a Blue Zone was new to him.

"First, you have to realize that Nicoya, like all of Costa Rica, has the best public health system in Central America," he said. "We have good sewage systems, immunization programs, and clinics in almost every village. Nevertheless, in Costa Rica we also have one of the highest rates of stomach cancer in the world. Many people die from it. But for some reason, it's the cancer mortality rate that is much lower in Nicoya. Perhaps they are eating something—or not eating something?"

I asked if there was anything else. Luis demurred and threw me a side glance from the driver's seat. He's a top-notch scientist, but sometimes he reminded me of a shy schoolboy, like the smartest kid in class who often had the right answers but was afraid of showing off. "You know in Latin America, we take marriage very seriously. If you get married, there's great pressure to stay married your whole life. But here," he hesitated. "Well, men here have very liberal attitudes toward sex. They tend to have many sexual partners throughout life."

That theory was a new one for longevity.

FIRST FORAY

Luis, Michel, Tom, and I set up headquarters in a hotel outside the city of Nicoya. Over the next three days, we searched for every centenarian or person close to the 100-year milestone within a 20-mile radius. Most of the 21 people we identified were easy to locate. Some four or five among these had died recently, but this proportion was extraordinarily low considering the ages. Luis had addresses and often phone numbers; the centenarians usually lived in simple, rural villages in cement houses located on streets laid out in a grid. Invariably, we'd be welcomed in and offered a seat.

Then Michel would get to work verifying people's ages. Since there are relatively few demographers in his specialty—long-lived people—he has invented several of the techniques now widely used to verify ages. Here, he asked to look at the birthday on the national identification card (*cedula*). Then he asked a series of questions about the parents' and children's ages to make sure they coincided. (For example, a 100-year-old woman who claimed that she had a 30-year-old son indicated a problem.) He'd take notes, including birth and death dates of all relatives and the numbers on their national identification cards.

Then I asked some questions about diet, lifestyle, and life history. Having interviewed more than a hundred centenarians in other places around the world, I wanted to get a general impression of how people aged here compared to people in Okinawa, Sardinia, and Loma Linda, California.

As a whole, the people I met in Nicoya seemed sharper and more active than anywhere else. They all believed in God, seemed to have a strong work ethic (like the Okinawans), and possessed a zeal for family second only to the Sardinians. All of them seemed to be taking advantage of Costa Rica's excellent public health system, receiving vaccinations, and using local clinics whenever necessary. Their diet consisted largely of corn, beans, pork, garden vegetables, and an abundance of fruit (papaya, mango, chico zapote, oranges), much of it grown in and around their yards. And the majority of the men, and a small percentage of the women, admitted to having lovers besides their spouses. None of them were officially divorced. They just started living with someone else.

Among the centenarians we met during our first week:

• Rafael Ángel Leon Leon, 100, who lived at the top of the winding road in the hills above Nicoya.

Behind his basic wooden farmhouse, he still grew a garden, harvested his own corn and beans, and kept some livestock. He played the guitar, had an excellent voice, and true to Luis's observation, fancied himself a lady's man. He'd had several lovers before he finally married an Indian woman 40 years his junior—when he was 94.

• Ofelia Gómez Gómez, who lived with her daughter, son-in-law, and two grandchildren. At 102, she recited from memory a six-minute poem by Pablo Neruda.

• Francesca Castillo (Panchita), who was 99 the first time we met her, lived alone outside Hojancha, cut her own wood, and twice a week walked a mile into town (once to attend Sunday Mass and a second time to go to the market).

At week's end, we gathered all of our notes and returned to San José. Michel was very excited. We all agreed that, along with Sardinia's Blue Zone, this was one of the most extraordinary regions of longevity we'd seen. Before we announced our findings and committed to a more thorough research trip, however, Michel had one more task to satisfy his scientific skepticism. We had an appointment to descend into the basement of the national archives and double-check the age of each person we met in Nicoya.

The process involved crosschecking each subject's national identification number to the corresponding entry in a sequentially numbered ledger located in the archive. This sort of eye-glazing detail made me hate science as a kid, but here we were about to enter the tomb containing the "Holy Grail" of data. It took about an hour to get the necessary clearance, and then two functionaries led us to the basement of the building where the vital information of every person born in Costa Rica since 1888 is recorded in floor-to-ceiling rows of leather-bound books. Hours later, after we checked every name in the ledger and the corresponding birth certificate, Michel emerged from the archives into the late afternoon like Rocky Balboa.

"This is a great success!" he announced triumphantly. There was no ambiguity: Nicoya, Costa Rica, was—along with the Sardinian Blue Zone—an area with one of the longest-lived populations in the world.

But the question remained: Why?

RETURN TO NICOYA

When Michel and I returned to Nicoya in January 2007, we came armed with a plan and a team of experts. We set up our headquarters at Dorati Lodge, a cluster of rough-planked, tin-roofed dormitories on the forested edge of Hojancha, near Nicoya's core. Run by a doting family, the lodge was close to villagers living a traditional life-style and small enough that we could take over the com-pound. It came with its own troupe of howler monkeys and was just down the road from a centenarian. (Michel had spent much of the intervening five months studying maps and spreadsheets to better define Costa Rica's Blue Zone and found that it was located only in the midsec-tion of the peninsula, excluding both the southern and northern extremes and the Pacific Coast.)

The family gave us permission to install a high-speed wireless Internet connection and to convert the open-aired dining pavilion—a sheltered, bush-flanked slab of cement—into a longevity laboratory and video-production house. Each night we'd create reports and short videos for an online audience who'd track our progress and offer their insights.

Our first recruit was Dr. Gianni Pes, who discovered with Michel the first Blue Zone in Sardinia. Like Michel, he was a careful, skeptical scientist, not inclined to jump to conclusions. He brought a suitcase full of academic papers and the Spanish translation of a survey he had used to assess Sardinian centenarians. He also had developed several hypotheses from other long-lived populations around the world that he wanted to test. Among them, Gianni observed that the more daughters a man has, the longer he lives; that people born in winter seem to live longer than those born in the summer, and that people who think they're going to live longer actually do. Would these findings also hold true in Nicoya?

Luis Rosero-Bixby (who could not join us because of a minor surgery) connected us with Dr. Elizabeth Lopez, a former psychologist with the World Bank who specialized in well-being. She developed a simple test that would help

> ## SLEEP TIGHT
>
> Getting enough sleep keeps the immune system functioning smoothly, decreases the risk of heart attack, and recharges the brain. Adults both young and old need between 7 to 9 hours per night. To help get it, go to bed at the same time every night and wake up the same time each morning; keep your bedroom dark, quiet, and cool; and use a comfortable mattress and pillows.

us determine levels of self-esteem, sense of purpose, and well-being among Nicoya's centenarians.

Eliza Thomas, a California-based researcher and health writer, would focus on identifying components of the Nicoya diet. Atlanta-based Sabriya Rice, a CNN producer and fluent Spanish speaker, would help with interviews. We also had a film and photography crew that included Damian Petrou, Tom Adair, Joseph Van Harken, Nick Buettner, and Gianluca Colla to help document our findings. Jorge Vindas, a Costa Rican who had interviewed many of Nicoya's centenarians as part of Rosero-Bixby's research team, would join us as a facilitator and translator.

After dinner the first night, I gathered the team and presented the plan. They sat at a long table, with the remnants of our bean, chicken, and tortilla dinner in front of them and listened. "We'll have two teams," I began. "Michel and Gianni will lead one team. Their goal will be to find and interview as many of Nicoya's elderly as possible. As results come in, they will report back to the rest of us on their general condition, what they eat, and any other observations that occur with regularity."

"You must know," Michel interrupted, his skin already pink from just a day of tropical sun. "Finding these things is a very slow process. You can't expect results right away—and maybe no results at all." Gianni, who was 48, with a neatly trimmed moustache, glasses, brown hair, and an air of shy intelligence, nodded in agreement.

"Thank you," I replied, meaning it. Though Michel's thoroughness was inconvenient, I knew that his findings would pass muster with other researchers. "The rest of us will be charged with finding people who fit the longevity profile and interviewing them to get their stories."

"Where do we start?" Sabriya asked.

"Here's what we know," I replied, not sure where I was going with my answer. Before arriving in Nicoya, Michel, Gianni, and I had spent several days at the University of San José talking to geologists, historians, epidemiologists, geneticists, and medical doctors. They had given us some strong ideas.

"Nicoyans have lived in relative isolation for the last four centuries, so their culture has developed differently from the rest of Costa Rica. But they are not genetically different. They have the country's lowest rate of cancer, so they're doing something here that protects them. And this part of Costa Rica has a strong indigenous influence. Finally, Gianni found a geology atlas at the university that showed the peninsula's unique physical feature," I said, noticing that the team was staring at me blankly. "Okay," I said, taking a step back. As a team, we needed to walk before we ran. "We'll give you names and addresses of centenarians. Why don't you take a couple of days to go out and get to know some of them?"

I took my own advice, and during the first week randomly chose a few of the dozen or so centenarians we had on our list to meet. On day three, I decided to visit 101-year-old Don Faustino, mainly because I had heard that he routinely took a bus each Saturday to Santa Cruz for market day, where every week he bought the exact same items on a grocery list.

Tomorrow was Saturday.

THE FIRST CLUES

The sun was just rising over the prickly pastures outside the village of Veintisiete de Abril when I arrived at the three-room wooden house that Don Faustino shared with his grandson Jorge and his 15-year-old great-grandson, Marco. They were already outside, waiting for the 7:30 a.m. bus to take them to town.

Don Faustino, who gets up at four on market day to get ready, wore a short-sleeve plaid shirt, floppy hat, and pants rolled up above his gnarled sandaled feet and sinewy ankles. He nodded an indifferent "yes" when I asked if I could join him on his trip to the market. Moments later, an old Greyhound pulled up with a hissing belch, and the doors swung open. We nudged onto a packed bus, where the temperature was considerably warmer than 98.6 degrees. Don Faustino's eyes met mine and he nodded again, this time as if to say, "You wanted to come."

We disembarked a block away from the farmers market where Faustino struck off on a serious quest for perfect red peppers and *platanos*. He touched perhaps a dozen vegetables before he bought six of each, paying with a few tattered bills. Next stop: the butcher.

"Why don't we just buy it here?" I asked Jorge, a portly, middle-aged man with a round, affable face. There were rows of stalls selling meat. "He has his rituals he doesn't like to break," Jorge sighed, mopping a sweaty forehead with a dirty rag. The tropical midmorning sun was already hot. "Moreover, the butchers are his friends."

So we walked another half mile, Faustino leading the way, resolutely shuffling down the street 20 feet ahead of

Every week, centenarian Don Faustino buys groceries for his family's Sunday meal.

us. I was thinking: "This guy is a machine." Most people Faustino's age couldn't get out of a chair.

"You don't forget your friends, do you Don Faustino?" shouted the butcher with a smile as we approached. Faustino shook his head "no," handing two-liter plastic bottles to the butcher to fill with liquefied lard. Then the butcher sliced off two slabs of pork from a dangling pig carcass and wrapped them in newspapers. Faustino paid with exact change and shuffled off again.

Down the block, in a general store stocked with canned goods and wilted produce, he bought sweet corn bread. This was not for him. "It's for my son," uttered Don Faustino in his soft voice. "It's his favorite." (This bit of thoughtfulness conjured images of a little kid receiving a much-awaited treat until Jorge reminded me that Faustino's son is 79.)

This time before paying Faustino made what seemed like an impulse buy—a calendar. He picked it up, flipped through it, paused to study the cover shot of a tranquil, forest stream, and tossed it on the checkout counter next to the bread. The title of the calendar: "The River of Life."

Back near the bus stop, we broke for an early lunch. Over a bowl of soup I asked Faustino about his life. He told me he loved to work, mostly because the fruits of his labor have provided for his family. For most of his career, he worked as a mule driver, hauling logs out of the forested hills and acting as a courier across the largely roadless Nicoya Peninsula. He also grew corn, beans, and vegetables to feed his wife and six children. "I also had two kids with a village girl," he said matter-of-factly, out of nowhere.

"Didn't your wife care?" I ask.

"I don't know," he shrugged. "We didn't talk about it." He went on to tell me, unapologetically, that he never gave the children his name nor did he ever support them. "How do I know if they're going to turn out okay?" he exclaimed as if explaining his actions.

On one hand, it seemed callous. On the other hand, Jorge Vindas, who was with me, had interviewed about 650 seniors in Nicoya and calculated that 75 percent of the men had had sex outside the marriage. He told me that Faustino was just being Nicoyan. Faustino fell into a pensive silence, then looked up, concluding with a non sequitur: "I've had a tranquil life."

On the bus ride home, I felt as though I was getting my first glimpse into the hardscrabble lives of these Nicoya men, these cowboys who worked hard and loved freely. Perhaps the isolation permitted a looser set of values to become the norm? Or perhaps this guy was a macho jerk who happened to survive a long time? I didn't think so.

Back at Veintisiete de Abril, I saw the truth. We stopped at Faustino's son's house, just down the street from his place, to deliver the sweet bread. Like most of the village houses, it was a dusty, three-room shack with walls of wooden planks, a tin roof, and chickens roaming freely in and out. There I witnessed Faustino's genetic cascade: his daughter, granddaughter, great-granddaughter, and great-great-grandson—five generations—all together in the same room. When Faustino walked in, his daughter Maria Jesus—a rotund 78-year-old—lit up and hugged her father. "Oh Papá, thank you!" she exclaimed. "You know I wait all week for this."

We moved into a dimly lit, rose-colored living room and settled into dusty sofas. Faustino's great-great-grandson, six-year-old Elias, curled up in his lap. Although Faustino was not part of the room's chatter, his presence there made the family seem complete now. I asked why Don Faustino made this weekly shopping expedition.

"He buys these ingredients for the exact same Sunday soup," replied Jorge. "After church, the family will all gather for dinner. He's been doing this for 40 years; it's the highlight of his week. We all look forward to it too, though we wouldn't mind a menu change once in a while."

"Why doesn't Maria Jesus make soup?" I asked.

"She would, but he won't let her. He still feels the need to provide."

This fact was interesting. A week earlier, I had met Dr. Xinia Fernández, a nutritionist who'd studied Nicoya. I had contacted her about dietary surveys of the peninsula but soon realized that over the years of talking with Nicoyans, she had gathered keen insights into their character as well.

"We notice that the most highly functioning people over 90 in Nicoya have a few common traits," she told me. "One of them is that they feel a strong sense of service to others or care for their family. We see that as soon as they lose this, the switch goes off. They die very quickly if they don't feel needed." Indeed, in every Blue Zone, centenarians possess a strong sense of purpose. In Okinawa it was *ikigai*—the reason to wake up in the morning. Here, said Fernández, the Costa Ricans called it *plan de vida*.

I looked over at Don Faustino. He was across the room with José sitting crossways on his lap. Absently or perhaps instinctually, he caressed the boy's bare foot as he stared out into the air. Was he thinking of the two kids he'd fathered 70 years ago? Or could this just be the look of tranquility? He had not been a perfect man by my definition, but he did possess a crystalline notion of priorities. He knew his plan de vida. They were seated in this room.

Here was a man who still pushed himself to rise at four every Saturday morning, travel to the market, and buy the food to provide for his family. Maybe it's the inertia of duty that keeps him going or the sense of caring and feeling good about doing something for others. Or perhaps it's the human imperative to feel needed that keeps the river of life running through Don Faustino.

CHOROTEGA FACTOR

The days of our expedition unfolded with satisfying progress. We'd wake each morning at seven and gather in the dining pavilion, where we'd take a breakfast of tropical fruit, beans, rice, and tortillas. Then we'd disperse like billiard balls after the break. Gianni, Michel, and Elizabeth struck out to do their interviews; their goal was nine a day. Eliza, Sabriya, and I would split up the video crews and photographers to track down leads. Just before sunset, the team would return, heat-flushed and dusty from the day, but also flush with some new discovery, and converge on the dining pavilion, which now was a sea of laptop computers, cables, and power cords, with a cooler in the corner.

After dinner, each team member shared their findings with the rest of the group. We hung a bed sheet from the rafters and took turns presenting our findings, using a projector. The video team made a three-minute Internet movie of one story that highlighted a component of Nicoyan longevity. Then Gianni and Michel took their turns, usually flashing a series of pie charts and bar graphs that revealed what their interviewees were eating and how they performed on physical tests.

By the end of the first week, Gianni and Michel had interviewed about 20 of the oldest Nicoyans. They asked them about diet and hours of sleep. They took blood pressure, pulse rate, and put each subject through short physical tests. At first blush, the Nicoyans were an impressive bunch. For Michel, a hypothesis was taking shape: "We're seeing a Chorotega factor," he said. "I think this is something very important."

He was referring to the Chorotega Indians who thrived in Nicoya before the Spanish arrived in 1522. They lived in simple thatched-hut communities ringed by vegetable gardens, fruit trees, and, farther out, *milpas* (fields of corn and beans). There, using slash-and-burn agriculture, they grew maize and many kinds of beans. They were mostly subsistence farmers, but they also depended on wild game and fish. Like other Mesoamerican cultures, the Chorotega were religious and apparently lived low-stress lives. And according to him, their influence was still alive.

"You see it over and over again; these people have Chorotega ancestry," he said, making a characteristic chopping motion with his hand. "These people are eating the same food and living with relatively little stress. But of course, we need to do more study. . . ."

Meanwhile, Eliza and Jorge had spent the day with 91-year-old Aureliano Rosales in the village of Santa Ana, and they were prepared to report on what they'd discovered. Eliza is a 27-year-old science writer from California with the wholesome good looks of a farmer's daughter. As a writer and editor of health magazines, she is familiar with every fad diet and "health supplement" that has come along in the past decade.

The countless stories she had researched on health and wellness had galvanized her belief in the subtle, long-term power of eating unprocessed foods. For this trip, she'd done several months' research into the nutrient values of fruits and vegetables that we'd expect to encounter in Nicoya. So when she heard about Aureliano's yard, which Jorge had described as "a private Garden of Eden," Eliza leaped at the opportunity to meet him.

Eliza stood up with a stack of notes and began clicking through a series of images. As we saw pictures of Santa Ana, Aureliano's wood-plank house, his family and garden, she narrated his story. Aureliano's life and that of Don Faustino were strikingly similar. Both lived in small villages, each with a central plaza and surrounded by vast cow pastures that dissolved into rolling hills on the horizon. An early life of hardship had tempered them into strong men with a zest for work (Aureliano's parents had died, leaving him an orphan). Both worked as mule drivers and subsistence farmers. They boasted of having raised big families but were quick to admit to the romantic adventures of their youths.

Aureliano, a decade younger than Faustino, was still an ebullient, sturdy man with an enviable head of white hair

Freshly cooked tortillas are a regular part of the Nicoyan diet.

and a contagious belly laugh. He shared a two-room house with his wife, daughter, granddaughter, and nephew, and he still lived a traditional Nicoyan lifestyle.

"Aureliano led me into the garden behind his home with a machete in hand, whacking weeds, pruning trees, and chasing away vermin. The garden had dozens of fruit trees," she said, clicking through images of banana, lemon, and orange trees. "But then he showed me these wildly exotic fruits." There was the *marañón,* a red-orange fruit five times richer in vitamin C than oranges; *anona,* which looks like a misshapen, thick-skinned pear known to have selective toxicity against various types of cancer cells; and wild ginger, a great source of vitamin B_6, magnesium, and manganese. "All of these are antioxidant powerhouses associated with disease prevention and longer life," Eliza informed us.

"At the edge of the garden, Aureliano showed me the corn and bean field he tends and harvests twice a year. He showed me how the low-growing beans provide moisture-capturing ground cover and fix nitrogen in the soil and, in turn, how the corn provides stalks for the bean vines to climb.

"He took me back to the house where he showed me how the corn is dried, then soaked in water with ash and lime, to loosen the tough outer skin," Eliza explained. "Then his wife demonstrated how the corn is ground, made into dough, patted into tortillas, and served with beans and squash. This creates the foundation of perhaps the best longevity diet the world has ever known," Eliza said with some excitement. "This food combination is rich in complex carbohydrates, protein, calcium, and niacin. Recent research shows that in high doses maize can reduce bad cholesterol and augment good cholesterol."

Eliza looked up at us and shuffled her papers. "In San José I interviewed Professor Leonardo Mata, who proposed that the most significant component of this food triangle is maize (corn). Here, the fact that they use lime—which is calcium hydroxide—to cook the kernels makes all the difference. It infuses the grain with a high concentration of calcium greater than in untreated maize and most other foods, and unlocks certain amino acids for the body to absorb. Nicoyans call the resulting maize dough *maíz nixquezado*.

"Mata has been studying diets in Guatemala where people prepare corn in that manner," she continued. "Populations consuming maíz nixquezado (maíz nixtamalizado in Guatemala, Mexico, and other countries) appear to have very low rates of rickets and also suffer fewer fractured bones and hips. This might also explain why Aureliano was so robust and strong while living all his life in relative poverty. So his fresh, organic, nutrient-rich diet with nixquezado explains why this 91-year-old looks like he's 60."

She ended the talk with a slide of Aureliano kissing his smiling wife on the cheek. We had only been there a week, and it seemed like Eliza had found Nicoya's longevity secret in Aureliano's garden.

"Unfortunately, it's not so simple," interjected Gianni, who had been listening quietly. "We don't know how much of the foods that grow in his garden Aureliano has eaten throughout his entire life, and we don't know what else he has also eaten. It could be something you didn't even see. In fact, we can associate his longevity more easily to the fact that Aureliano lives with a daughter than what he grows in his garden." Then Gianni went on to describe a study done in rural Poland showing that every daughter increases a man's life expectancy by 75 weeks.

Eliza, the good-natured soul that she is, accepted the feedback and asked Gianni how we could tease out Nicoya's diet of longevity.

"You must begin by finding out what most centenarians have eaten for most of their lives," he replied.

Over the next week, we worked to draw a bead on the Nicoyan diet. Gianni and Michel's interviews yielded some

---

ORANGES

One of the most popular fruits consumed in Costa Rica's Blue Zone is one of the easiest to find at home—the orange. Suppliers of vitamin C, fiber, potassium, and folate, oranges have been credited with helping prevent heart disease, cancer, and stroke.

answers. They asked centenarians what they ate and heard "beans, rice, tortillas, and fruit" over and over. This was an imperfect source: People's memories can be inaccurate. (Quick—what did you eat for lunch a week ago Tuesday?) But they would offer clues.

Eliza had a better source. In a pile of academic papers given to her by Dr. Fernández, she had found three national nutritional surveys dated 1969, 1978, and 1982. Costa Rica's Minister of Public Health, we discovered, periodically surveyed people in each of the country's five rural regions to determine what and how much they ate.

The surveys focused on one town the ministry considered representative of the region and measured how many grams of each of 20 foodstuffs people ate—things like dairy products, grains, tubers, and fruits. In 1969 the survey focused on Veintisiete de Abril (Don Faustino's town) and in 1978, the survey took place in La Mansión—two miles from the Dorati Lodge!

I had another source. In preparation for the trip, I had found a University of California Press report from 1958 entitled *Nicoya: A Cultural Geography* by a Berkeley anthropologist named Philip Wagner. I had a photocopy of it in my luggage. Eliza and I thumbed through it, and on page 241 we struck research gold—a chapter that described a day in the life of an average Nicoyan—*50 years ago*. It read:

> *The day of the country people begins before sunrise, when the women rise to prepare coffee. The family meets about dawn to take its cup of black coffee, or coffee with milk, heavily sweetened, and perhaps to eat a cold tortilla. The*

time from dawn to eight o'clock is for chores and beginning the day's work. At eight there may be a complete breakfast with rice and beans and eggs. In seasons of heavy work the men take with them to the fields tortillas with gallo pinto *(rice and beans fried in pork fat). Work may end on very hot days at twelve, or at two in the afternoon. The workers come home from the fields or the woods and wait about an hour for their meal.*

The midday meal often begins with a pot of soup in which there are a few bits of meat, fat, boiled plantains, tisquisque [sic] *or yuca, and perhaps a few greens. After the soup come rice and beans, usually accompanied by fried eggs. On occasion there may also be some vegetable: pipián or ayote* (Cucurbita moschata) *or calabaza* (Lagenaria), *cabbage, the flower of piñuela* (Bromelia pinguin) *or some other wild product. Meat sometimes appears on even the poorest table, and there is usually* cuajada, *a milk curd. Tortillas come with this meal and afterward the men sip heavily sweetened black coffee, made from local berries or from the mashed seeds of* ñanjú (Hibiscus esculentus).

The evening meal is simpler, since the custom is to spend the afternoon in idleness, and appetites are less hearty. Rice and beans, tortillas, and perhaps eggs are served just at dusk.

Wagner also made detailed sketches of gardens, showing more than 40 different edible plant species (many still growing today in Aureliano's garden), but highlighted yucca, tiquisque, papaya, bananas, yams, and nanpí as mainstays of the local diet. Nicoayans also ate a variety of exotic forest fruits like jobo (*Spondias*), guayabo (*Psidium guajava*), caimito (*Chrysophyllum cainito*), and papaturro (*Coccoloba*).

So what's the upshot? One night after dinner, I went back to my cabin to sort things out. I sat cross-legged on the wood plank floor and spread out my papers in a huge semicircle. I had Wagner's accounts, the dietary surveys, academic papers, and my notes on diet in Costa Rica.

A few characteristics of Nicoya's diet stood out. Like the people in most other Blue Zones, Nicoyans ate the emblematic low-calorie, low-fat, plant-based diet, rich in legumes. But unlike other Blue Zones, the Nicoyan diet featured portions of corn tortillas at almost every meal and huge quantities of tropical fruit. Sweet lemon (*Citrus limetta*), orange (*Citrus sinensis*), and a banana variety are the most common fruits throughout most of the year in Nicoya. Mata's explanation of nixtamale seems to explain the connection between maize and longevity, but how do fruits contribute to longevity?

Luis had told me on our first trip that Nicoya had lower rates of stomach cancer—a major killer in the country. In San José, I had tracked down Professor Rafaela Sierra, a Spanish-born epidemiologist from the Institute de Investigaciones en Salud (INISA) to see if she could tell me why. Sierra explained that stomach cancer is closely linked to the *Helicobacter pylori* (Hp) bacteria. She did a study comparing people living in the area with the highest incidence of stomach cancer (near San José) and the lowest rate of stomach cancer (Nicoya) to try to identify the cause. "I found that school kids in both places were equally infected with

Hp," she said. "So we knew that it wasn't just the bacteria. The cancer comes from how people react to it, and there are many things that mix."

When I asked her to venture a guess as to why the incidence of cancer is so much lower in Nicoya, she hedged. "We don't know anything for sure, but I can tell that the populations of both zones were dominated by peasants. Near San José, they tended to be migrant workers, moving around as seasonal work presented itself. In Nicoya, the peasants tended to be sedentary. We know that the big landowner often let them grow a garden, which gave them access to fresh vegetables.

"We also believe that vitamin C and beta-carotenes may help prevent stomach cancer or at least subdue the effects of the Hp. You find these nutrients in fruits and vegetables like papaya, carrots, *calabazas* (squash), oranges, pineapples."

I wasn't familiar with all the different kinds of fruits and vegetables available to Nicoyans, but all of our team's research and everything we had observed directly in the field pointed to Nicoyans being big consumers of fresh fruit. Our research also suggested that fruits' role in preventing stomach cancer seemed to offer a piece in Nicoya's longevity puzzle.

SOMETHING IN THE WATER?

Near the end of our second week in Nicoya, Gianni and Michel were putting a couple other pieces in place. In San José, they'd followed their noses into the basement of the university library, where they'd found an atlas illustrating different features in each of the country's regions. They were looking for idiosyncrasies, characteristics that distinguished Nicoya from the rest of Costa Rica. The atlas provided three leads: 1) Nicoya was the hottest and driest region; 2) It received more hours of sunlight on average during the year; and 3) The water percolating up from Nicoya's limestone bedrock was very different than other water in Costa Rica.

When Gianni matter-of-factly mentioned that Nicoya's water was special, it immediately conjured up notions of Juan Ponce de León and his legendary 16th-century search for a Fountain of Youth. Indeed, though Ponce de León's quest began in what is today Florida, the rumors originated with the Arawak Indians, who inhabited islands off Costa Rica's Caribbean coast. But how was Nicoya's water special?

"What the atlas showed, specifically, was the mineral content of the water," he replied. "It revealed that the water hardness, the calcium and magnesium content, was higher in Nicoya than anywhere else in Costa Rica." To confirm this, Gianni used a water-testing kit; in each of the 20 or so households where he conducted interviews, he tested the drinking water. The result: "The water had such high levels of calcium and magnesium that I had to dilute it by 50 percent with distilled water just so that I could test it."

Gianni then calculated that if the average Nicoyan consumed (through drinking, cooking, or making coffee) six liters of water daily, he or she would ingest a gram a day—for most people their daily requirement of calcium. And how does calcium-rich water explain longevity?

Gianni pushed his plate aside to make room for his laptop. He opened a file and showed me a 2004 World Health

Panchita, 100 years old, rests after a day spent cooking, splitting logs, and using a machete to clear brush from her yard.

Organization paper that examined a host of studies over the past 50 years. Populations with hard water, it found, have up to 25 percent fewer deaths from heart disease than populations with soft water. I asked him how he explains this. "The heart is a muscle, and all muscle contractions depend on calcium," he said. "Inadequate calcium means weak muscles—including the heart. Old people often have too little calcium in their bodies. So having extra-hard water may help keep Nicoyans' hearts strong for longer."

Also, calcium is important for bones, he continued, looking at me over his glasses. I could tell he was making a good-natured effort to simplify the explanation. "Our bones are constantly losing mass and building it back up. When we are young, our bones build faster than they deteriorate. After we are about 40, the reverse occurs, and we lose bone faster than we build it. Calcium may help slow that loss. Lastly, loss of bone strength can also make older people more susceptible to hip fractures, which is one of the leading causes of death for older people. Adequate levels of calcium might explain why Nicoyans are avoiding risks that kill other populations. "But (there was always a "but" with the ever-cautious Gianni) before we can say for sure that Nicoya's water is part of the explanation of its longevity, we must do more tests."

SAFETY NETS

One night toward the end of the expedition, it was my turn to stand up after dinner and present a report to the team. I told the story of Abuela Panchita, the 100-year-old woman whose 80-year-old son, Tommy, biked to see her each day. In many ways she represented everything that I'd learned so far about Nicoyan longevity: Successful centenarians here were religious, family oriented, unconcerned with money, flexible but ultimately decisive, and consummately likable. I clicked through pictures of Panchita chopping wood, clearing bush with a machete, and walking through town in her bright pink dress and carnival beads. I told everyone that of the 200 centenarians I have interviewed around the world, Panchita was the most extraordinary. "You've got to introduce me to her," said Elizabeth Lopez after my talk. Elizabeth was our team psychologist, who specialized in well-being. "I've done 20 interviews so far, and I've never seen anyone like her."

Panchita lived just a few hundred meters from the Dorati Lodge, so Elizabeth and I left the next morning on foot. We walked past the howler monkeys in the mango trees overhead, out the driveway, and into the village of Hojancha.

There, mothers walked their white-shirted, uniformed kids to school, and old men swung on hammocks on porches. Something about the Costa Rican morning—the ever-present soundtrack of *sabanero* music, the faint aroma of tortillas roasting, or the radiant sunshine instantly makes me happy.

Elizabeth had come to us by way of Luis. She had read an article in the Costa Rican newspapers about the Blue Zone project, and, having recently retired from the World Bank, she was looking to get involved with something, her new plan de vida. So Luis put her in touch with me, knowing our project was short on local expertise. When we met, I immediately recognized her as a godsend. A Costa Rican native, she spoke fluent Spanish, and thus served as a perfect liaison between Gianni and Michel (who did not speak Spanish) and the interviewees. Moreover, she could help them expand their questionnaire to include a way to measure psychological factors—like happiness and faith levels in long-lived people. As we walked, I asked her what she was finding.

CALCIUM

Nicoya's drinking water is richly packed with calcium, giving the locals an easy supply of this important mineral. Calcium is the most abundant mineral found in the human body and vital to keeping bones strong. For those without calcium-rich water, yogurt, milk, and cheese are great natural sources of calcium. For dairy alternatives, try sardines, kale, and broccoli.

"Dan, these people are so incredible," she answered enthusiastically. "They are *so* positive and so devoted to their families. All but one of the 33 Nicoyans we have met live with their family." Elizabeth was looking at me, gesticulating as we walked. "They have a wonderful support network. They also tend to have a large number of visitors that they receive almost every afternoon, which is both a physical and psychological safety net."

We walked over the same bridge that Tommy and I had biked over a few days earlier. Toward the edge of the village, the houses were simpler, with pole walls, tin roofs, and small yards. The poorest and most "Indian" families lived on the outskirts of Hojancha, where the rectangular street pattern gave way to winding paths and roads that often followed the banks of streams. In one such house Panchita lived.

As we turned down Panchita's drive, I asked Elizabeth if she'd observed anything else. She thought a moment and replied. "Yes, two things. The first is—and this is just my impression—but I think that men live with less stress than women do. They don't tend to worry about the children . . . and we've all seen how common it is for men to have lovers outside the marriage."

GOD WILL PROVIDE

As we neared Panchita's house, we loudly called out for her. She pushed open a wooden shutter and, when she recognized me, raised her hands in unmitigated joy. She hurried out into the courtyard to hug both of us. "Panchita," I said in a raised voice (she is partially deaf and blind). "This is

Elizabeth. She's a scientist from San José. She wants to visit with you."

"Oooo!" she whooped. "Of course. Come sit down." She was wearing a festively frilly dress like the one she wore the first time I met her, but this time it was green instead of pink. Long, green earrings dangled from her ears, and she had pulled her gray-tinged hair back with a rhinestone-studded comb. She led us to the two wooden benches that line her porch.

Elizabeth quickly established a rapport with Panchita and asked her about her life as a child and young woman. Panchita told her that she was a descendant of a Cuban revolutionary hero and that she'd had a beautiful childhood.

"In those days, there were no roads in Nicoya. My father owned a guesthouse, and occasionally mule trains would come by. I woke up at three each morning to make coffee and tortillas for the men who stayed overnight. I took care of my parents," she said. Then, turning to me, she scolded congenially, "It's like this, Papi. Those who honor their parents are rewarded by God."

Panchita eluded our direct questions about her husband and the father of her children. From her answers, we were able to ascertain that she did raise her children mostly by herself. They all lived with Panchita's parents until they died, then Panchita inherited their farm. There, the family grew most of their own food. When they needed salt or sugar, Panchita would walk the 18 miles to town and back to get it.

"Life was hard those days, Papi." She always called me Papi. "You can no longer cook me with a little bit of water," she said, referring to her age and that old tough meat takes longer to cook. As she spoke, she waved a finger at me. Spittle flew from her mouth, and her dangling legs kicked reflexively.

At one point during our conversation, her normally festive demeanor turned serious. She put her hand on my arm; she had an endearing habit of gently, instinctually touching people to make a point. I looked down at her long, smooth fingers and neatly trimmed nails. She wore a dented silver band on her ring finger.

"They killed my son," she said fixing me with a brown-eyed gaze. The lines in her lightly creased face reflected the sadness of a 50-year-old tragedy. "When he was a beautiful 20-year-old man, he got into a stupid fight with a friend, and he killed my son." She sat silent for a minute, her dangling legs still swinging back and forth. "God does everything for a reason, though," she resumed brightly. "I am a blessed woman today."

Elizabeth turned to me and smiled in a way that said: "See what I mean?"

Another telling story: One day when Panchita was about 70, she was bathing in a river just south of town. She heard some rustling in the bushes and then noticed a man was watching her. "I rapidly put on my clothes and picked up a stick," she said, swinging an imaginary limb over her head. "And then I chased him down and almost beat him to death." She finished the story and her mood turned wistful. "Oh Papi," she said finally. "That was a very bad thing. I had to ask the priest for forgiveness. But still, God blesses me."

Late in the morning, the neighbor boy, 10-year-old Luis, arrived, as he does each day, to help Panchita catch her free-range chickens and put them in the coop. Later, her 31-year-old neighbor, Carmen Gómez, stopped by to help Panchita sweep her floors. "I don't come here because I have to," she told me when I asked. "Panchita has a way of making my day happier. Everyone in Hojancha loves her."

At noon, Panchita told me it was time to make lunch. Elizabeth and I followed her into her simple house: a room with a table and chairs; a bedroom with tattered sheets and blankets, and pictures of Jesus on the wall; and a kitchen, where she led us.

The kitchen was simple and pleasant: a small, well-lit room with two windows that opened to the yard, a small pantry, a wood counter, a soapstone sink with running water, and a small refrigerator. A bowl of bananas and papaya sat on the counter for easy access, and everything else—beans, onions, garlic, greens, corn, which all required preparation—remained out of sight. Cheese and tomatoes were the only things in her refrigerator. She had no pack- aged or processed foods; everything required preparation except the fresh fruit.

Panchita still cooked on a wood-burning *fogón*, the tra- ditional Chorotega clay oven that looks a bit like a wide- mouthed volcano atop a wooden platform. Elizabeth and I silently watched as Panchita prepared the food. She moved slowly and deliberately, heating up beans and seasoning them with garlic and onions. From an earthen pot she scooped out grayish corn that had been soaking in lime overnight, rinsed the kernels and ground them with a hand crank into dough.

She patted out tortillas and roasted them over the open fire. She melted a dollop of lard on an iron griddle and fried eggs. Finally she cut paper-thin slices of fresh cheese—an impressive feat given the fact she could barely see the cheese, much less her fingers.

In about 30 minutes she presented us with lunch—small portions of beans, corn tortillas, and one egg on a small plate. The serving looked huge, but it amounted to about half of what you'd get if you ordered the breakfast special at your local diner. "Food gives life!" she shouted and told us to sit down and eat. I felt both humbled and privileged to be served a meal prepared by a centenarian. But after much polite arguing, I managed to convince Panchita that she and Luis should eat the meal. Elizabeth and I had lunch waiting for us back at the Dorati Lodge.

On the walk back, I told Elizabeth that I agreed with her observations regarding faith and longevity. Panchita's faith was amazing—her unwavering belief that no matter how bad things got, God would take care of everything. Indeed, thinking back, I realized that most of the 200 centenarians I had met believed in a similar guiding power. The Seventh-day Adventist faith was rooted in a strong faith tradition; Okinawan elders believed that their deceased ancestors watched over them; and Sardinians were devout Catholics. Panchita was yet another example, it seemed, of the power of faith. I asked Elizabeth if faith really has a profound impact on longevity.

"Absolutely," she said. "When Gianni and I were doing our interviews, we noticed that when you ask the most highly functioning seniors how they are, they always say, 'I feel good . . . thanks to God.' Yet they may be blind, deaf, and their bones hurt. Psychologists call this an external locus of control. In other words, they tend to relinquish control of their lives to God. The fact that God is in control of their lives relieves any economic, spiritual, or well-being anxiety they might otherwise have. They go through life with the peaceful certitude that someone is looking out for them." I had heard of a study that echoed their findings: In this study, participants who attended religious services about once a month or more had up to a 35 percent reduced risk of death for the next 7 years.

Later in the afternoon, Elizabeth visited Panchita again and asked her some more questions. "I wasn't going to share this with you," she told me that night at dinner when I asked her about her experience with Panchita. "I was alone with this lovely, magical person. She doesn't live in a nice home. She's so poor yet so satisfied with what she has. There was a total acceptance. But I wanted to help her anyway. So, I handed her twenty dollars."

"And . . . what happened?" I asked.

"She told me, 'I had no money to buy food. But I knew God would provide,' squeezing my arm. 'And now He has.'"

DOCTORS' DIAGNOSIS

On the next to last night of the expedition, Gianni and Michel stood up in front of the team to present their final findings. They had prepared one of their elaborate pie-chart-and-graph PowerPoint presentations with a soundtrack. The two had been inseparable during the trip. They always sat at the same table with their laptops open, discussing their work earnestly, always in French. They were the first ones up in the morning, they did interviews together, and ended up at the same table at night, where they consolidated their findings in a giant spreadsheet. They had both made sacrifices to join us in Costa Rica (it took months for Gianni to apply for permission from his university to get the time off) and they had both taken a chance that their efforts would bear fruit. But today they were excited, and standing before the team, they looked like two big kids on a particularly special day of show-and-tell. Michel began by flashing a screen that read:

Costa Rican Blue Zone
First Results of the Survey
Hojancha, 02/07/07

"We did not know if we'd be successful with our investigations," Michel began. "We managed to find all the oldest of the old people living in Nicoya, and among these, eight females and six males were centenarians. This is important because it is not a sample but rather an entire population!" he said triumphantly, making big circular gestures. "And we managed to verify all of their ages—this is very important—in either the local church baptismal record or in the municipal hall."

He flashed a series of slides that showed the characteristics of the Nicoyans he interviewed. "More than half were of Chorotegan descent. They tended to sleep about eight hours a day. They got maximum exposure to the sun. The long-lived women were likely to be the firstborn. And we also noticed that these people were extremely positive, which we can associate with their longevity."

Now Gianni took the floor. "In order to assess the Costa Rican centenarians, we decided to compare them with Sardinian centenarians, about whom we have good data," he began. "What is surprising is the greater fertility of the centenarians from Costa Rica in comparison to the Sardinians. This does not mean that these differences are necessarily related to longevity. They might simply reflect the family structure in the two populations," he said. Unlike Michel, Gianni's hands remained folded motionlessly in front of him. Gianni, like most true longevity scientists, was reluctant to make absolute statements. And besides, longevity is an exceedingly complex field.

COOKED CORN

Costa Ricans have eaten maize (corn) since the time of the Chorotega Indians. Lime (calcium hydroxide) is used to cook the kernels, which infuses them with higher concentrations of calcium. But just exposing corn to heat can also increase its nutritional value. A recent Cornell University study found that cooked corn contains higher levels of antioxidants than raw kernels.

"Gianni," I interjected. "I know that it's premature to make concrete conclusions, but you've studied centenarians for almost two decades. What's your gut feeling as to why these Nicoyans are doing so well and living so long?"

"One moment please. I'll show you," he replied. He displayed bar graphs that compared body weight, blood pressure, and the results of physical tests. The Costa Rican centenarians, as a rule, were taller, weighed less, and had slightly higher blood pressure.

The Nicoya Peninsula, he and Michel had concluded, was, in fact, among the places in the world with the longest-lived populations, and Nicoyans performed significantly better on physical tests. Together with the mountainous part of Sardinia, Nicoya had to be considered as the second validated longevity Blue Zone.

Their trip to Costa Rica had been worth all the trouble, after all.

WHAT LITTLE IS SAVED

Before I left for home, I wanted to accomplish one more thing. Neither Costa Rica nor the once-isolated Nicoya had eluded progress. Each year more and more Americans were finding their way to Nicoya's coast, with its dependably dry, sunny winter and its dramatic coastline and white-sand beaches. Young families come to Tamarindo in the north for winter vacation; meanwhile the backpacking set heads farther south to beach towns like Montezuma for cheap rooms and excellent surf.

Along with the tourist dollars have come development and an influence on younger people. As with many of the Blue Zones, Nicoya's longevity culture is a disappearing phenomenon. I wanted to experience it as purely as I could before it was gone. Were there any more Nicoyans still living the authentic traditional lifestyle? And if so, could I find them? And could they offer any more insights into the region's incredible longevity?

Of course I had Wagner's manuscript to help me recognize the traditional lifestyle if I found it. In a chapter titled

"Settlement and Land Use," he offered some glimpses:

> *Men work slowly but steadily in the fields. The workday begins about six in the morning and runs until twelve or sometimes two in the afternoon. Few men work after two. Men may observe a break about nine in the morning to eat a little and to drink from the* nambiro *(water vessel). After the work is over for the day, they are content to sit about until dark, smoking, conversing, or napping.*

In the comments and conclusion section he wrote:

> *The individual lives in close relation with a world of plants and still turns directly to nature for the satisfaction of his wants. The subsistence economy, and the resources and techniques it employs, strongly resembles the Chorotegan pattern at the time of conquest. Although gods, rulers, tongues, and even races have succeeded one another, the old basic ways of life persist, and in the daily business of living, local forms and traditions maintain a stubborn continuity. . . .*

I shared this passage with our ever resourceful Costa Rican assistant, Jorge, and asked him if he could find anything like it. "Give me a day," he said. That night, he returned to the Dorati beaming. He'd spent an afternoon in Nicoya city talking to seniors, museum curators and domestic workers who typically come from the surrounding countryside. He'd gotten a lead from a policeman. "There's one place left," Jorge chirped in Spanish. "It's in the hills above the

This Nicoyan woman bakes tortillas and walks five miles to the village to sell them.

city of Nicoya. A place called Juan Díaz. We'll need to walk several hours."

The next day, following a map the policeman had drawn, we drove up a steep, rutted road that snaked into the forested hills along the peninsula's spine. At a roadside shack we parked and got out. Three little kids stared at us as we started walking along a footpath that led into the bush.

We hiked for about an hour and a half through a corridor of dense vegetation. The sun beat down and sweat soaked our clothes. Occasionally the jungle would open and we'd catch a sweeping view of Nicoya, spilling down the mountain to toast-brown flats that stretched all the way to the slate-blue line of the Pacific Ocean.

Perhaps an hour into our trek, we rounded a bend and came upon a two-room shack outside of which a shirtless man was fiercely chopping wood in a small jungle clearing. "Hello, I'm Dan Buettner," I said, extending my hand. "Do you know where we might find the village of Juan Díaz?" He put his sweaty, calloused hand in mine and let me shake it. He wore rolled-up trousers and rubber boots without socks; his ripped, tea-colored torso glistened with sweat. I guessed him to be about 60. "I'm Juvenil Muñoz, and you're in Juan Díaz," he replied flatly. I looked around. We were in a small jungle clearing. A flock of chickens, two cows, and an ox roamed freely. This was a town?

"May we stop for a visit?" I asked. Juvenil looked at Jorge, then back at me. He looked as if he thought we were playing a trick on him, but in the typical unquestioning character of the region, he replied, "Why not?" and dropped his ax. His house looked much like Panchita's—except it lacked running water and electricity. The kitchen looked pure Chorotega: a fogón, corn kernels soaking in a black earthen pot, hollowed gourds for drinking, a packed-dirt floor, a dog sniffing for crumbs. The second room was for both sleeping and storing feed.

He dutifully answered our questions: What do you eat? ("Beans, tortillas, fruit, and once a year, beef when I butcher a cow.") When do you go to bed? ("When the sun sets.") When do you wake up? ("When the sun rises.") He kept looking over at his ax, as if it were waiting for him. "I should finish my work," he said apologetically.

After three weeks in Nicoya, it seemed perfectly natural of me to ask to see his house, and natural that he'd say yes. But this was the first time I'd stopped to think about what I'd do if two sweaty strangers stopped at my house, interrupted my work, and started asking me inane questions. I'd probably call the police. "Of course," I said. "I'm sorry for the interruption."

Juvenil went back to his wood chopping, heaving his enormous ax in the air and bringing it down mightily. "By the way," I asked leaving. "How old are you?"

"I just completed my 90th year," he replied, not looking at me. Then he brought down his ax with a whoosh and thud that sent up plumes of woodchips.

A half hour further up the path we came to the second house. The scene looked exactly as Wagner had described Nicoya a half century ago: crude wood buildings, a raised chicken coop, a barn the size of a single-car garage, a mule-driven sugar cane press, and a small house of vertical planks. The tin roof provided the only hint that the 20th century had come and gone. An old man napped languidly in a hammock on the porch. Inside, we could hear talking and laughing.

"*Buenas tardes,*" Jorge called out. The talking stopped abruptly. An old portly woman wearing a housedress and an apron pulled aside the sheet covering the front door and looked out at us. "My name is Jorge Vinda and this is Dan Buettner, a journalist. Can we come in?"

"*Si, si,*" she said in a "*What are you standing out there for?*" sort of way. Inside it smelled like roasted corn and coffee. A small fire burned in the fogón. On a shelf, pottery jugs held drinking water. A calendar hung on one wall. A cool, dry breeze blew through the room from open shutters, sending curtains fluttering. A middle-aged couple sat at a wooden table staring at us with a bewildered look. Jorge introduced us and explained our project.

This was the farm of 91-year-old Ildifonso Zuniga and his wife, Segundina. Ildifonso was napping in a hammock on the porch; Segundina was hosting the guests. Ananias Baltoano and his wife, Aida—the middle-aged couple— had stopped by for a visit, which seemed normal enough until I learned that this visit had occasioned a two-hour, five-mile walk. They were, in the words of Segundina, neighbors. She told us to sit down, poured us a cup of coffee, and cut us giant pieces of sweet cornbread. We were now part of the family.

Instead of launching into interview mode, as I usually do, I just sipped my coffee and listened. The conversation moved languidly from the weather to the condition of Aida's 12-year-old daughter, who had broken her leg ("The poor little thing!" Segundina exclaimed.), to the birth of a neighbor boy, to the bean harvest. "We cut down bush and plow. I brought in over 50 kilos last week," said Ananias, who was a timid-sounding man with ruddy cheeks and jet-black hair. The conversation served no purpose, it seemed, other than to connect these friends and generate some idle entertainment. I wrote in my note-

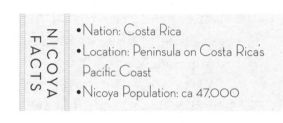

NICOYA FACTS
- Nation: Costa Rica
- Location: Peninsula on Costa Rica's Pacific Coast
- Nicoya Population: ca 47,000

book: "The room has a feeling of timeless congeniality."

After a quarter of an hour, I finally butted in. I told my hosts what we were doing with the Blue Zones and admitted that we'd come here because we'd heard that we could maybe see how people in Nicoya have lived for centuries. They nodded their heads. "For example, could you tell me how a typical day unfolds?"

"Well," said Aida, who wore a tattered brown dress and her gray-flecked hair under a scarf. She had the pink-blushed countenance of a face you'd see in a 16th-century Flemish oil painting. "We wake up when the sun rises. I make breakfast for the family and then Ananias takes our boys into the fields. Meanwhile the girls and I stay home and clean up. At noon or so, the men come back. We eat our 'strong' meal and then relax. We maybe visit someone like today or someone visits us. We have a light dinner. Usually we're in bed by 8:30," she concluded. "Since we don't have TV, there's not much for us to do after sunset."

Segundina poured us another cup of coffee, not bothering to ask if we wanted one. A cool breeze blew through, fluttering the curtains. Ananias drew in a deep breath as if taking a long drink of water on a hot day.

"And for food, what does one eat in Juan Díaz?"

"Beans and rice," replied Aida, and after a well-timed pause added: "Or rice and beans. It depends how we feel." Everyone laughed.

"We eat what God provides," Ananias said, now serious. "Eggs, rice, beans. Sometimes we kill a chicken."

"How often do you eat meat?"

"It depends how the pocketbook is," Ananias replied, patting his pants. "If it's good, maybe once or twice a week." Later I learned that their only annual cash income came from the couple hundred pounds of sugar the village press produced each year.

"We don't need much," Aida interjected, heading off an implication of poverty. "We're satisfied. . . ." Here Aida's answer trailed off and she fell silent for a moment. "You have to keep busy," she resumed, now answering a question I did not pose. "When people have too much time they get involved with vices. Here we have enough to do. We stay busy enough to keep the Devil away, but not so much that we get stressed. It's a clean, pure life."

"But don't you ever get bored?" I looked around. There was no TV, no radio, or electronic entertainment of any sort. I could not think of a single place I'd visited during the past decade—not the Sahara, the Amazon, the Congo—that didn't at least have satellite TV. "What do you do to entertain yourself?"

"I find a patch of shade and eat an orange," she answered, not skipping a beat.

Jorge and I hung out for another hour or so. For once, I decided to turn off the journalist mode and just experience the afternoon. We drank more cups of coffee and ate another piece of cornbread. I showed my new friends wallet photos of my family, and they swooned.

Soon talk gravitated back to village life, the forthcoming rains in May, the harvest later on, and then a community celebration in September. I resisted the urge to look at my watch or to think about what I wanted to write about this experience or about checking my email. I finally gave up and said I had to leave.

"Already?" said Segundina. "You should stay for dinner."

Jorge and I made our way back toward the Dorati Lodge. It was now late afternoon and the cooler, downhill hike felt good. We jumped into in a van that was blast-furnace hot from the afternoon sun. Jorge turned the key, and the engine started with a jolting roar. A CD began to blare. I quickly turned it off and rolled down the windows. We coasted pleasantly back down the hill until we reached the main road. There we accelerated abruptly and joined the stream of cars headed south.

We passed a string of truck-stop bars that doubled as brothels, the Nuevo Tempique Hotel (*aire acondicionado!*), the cement factory with its mountainside strip mine, and, near Nicoya city, the "Gran Apertura" of the Burger King—the peninsula's first. I regarded it all with mild disgust. "Enduring" just one quotidian afternoon in Juan Díaz had recalibrated my tolerance for our world's din and made the idea of quietly eating an orange in the shade of a tree seem genuinely appealing.

I now wish I had accepted Segundina's invitation and stayed for rice, beans, and more simple conversation. I imagined stepping outside and looking up into the star-spangled cobalt night, unsullied by light pollution, having a good night's sleep in a simple bed, and waking with the sun for a day of hard work as Ananias or Juvenil or Don Faustino have for their entire lives.

I reached into my courier bag for a bottle of water and saw Wagner's article, which I'd bound with a green cover. I plucked it out and flipped through it. On the last pages, I read something I'd highlighted earlier:

> *The commercial revolution is coming to Nicoya, and the growing availability and prestige of commercial products are creating a new demand for money which can be obtained only by producing crops or manufactured goods for sale. The small subsistence farmer, fully occupied with the task of securing a bare livelihood, possesses neither the land necessary for growing commercial crops, the time to tend them, nor the capital to finance the technological improvements essential for successful commercial production . . . Little attention has been paid to the possibilities of extending the cultivation of the crops of the Nicoyan garden, which require little land and slight care and yield abundantly . . . When the old ways disappear, as perhaps they must, it is regrettable that so little is saved from them, so that those who practice them suffer the penalty of obsolescence, as the poor folk of a new and more efficient world.*

For nine months, our team of scientists and researchers studied the amazing people of this tiny peninsula in northern Costa Rica. Their choices and traditions revealed new insights into the mysteries of longevity. Their unique physical environment coupled with their time-honored lifestyle have come together to yield the world's healthiest, longest-lived people, the inhabitants of a new, true Blue Zone.

COSTA RICA'S LONGEVITY SECRETS

Try these common practices from Costa Rica's Blue Zone.

Have a plan de vida.
Successful centenarians have a strong sense of purpose. They feel needed and want to contribute to a greater good.

Drink hard water.
Nicoyan water has the country's highest calcium content, perhaps explaining the lower rates of heart disease, as well as stronger bones and fewer hip fractures.

Keep a focus on family.
Nicoyan centenarians tend to live with their families, and children or grandchildren provide support and sense of purpose and belonging.

Eat a light dinner.
Eating fewer calories appears to be one of the surest ways to add years to your life. Nicoyans eat a light dinner early in the evening.

Maintain social networks.
Nicoyan centenarians get frequent visits from neighbors. They know how to listen, laugh, and appreciate what they have.

Keep hard at work.
Centenarians seem to have enjoyed physical work all of their lives. They find joy in everyday physical chores.

Get some sensible sun.

Nicoyans regularly take in the sunshine, which helps their bodies produce vitamin D for strong bones and healthy body function. Vitamin D deficiency is associated with a host of problems, such as osteoporosis and heart disease, but regular, "smart" sun exposure (about 15 minutes on the legs and arms) can help supplement your diet and make sure you're getting enough of this vital nutrient.

Embrace a common history.

Modern Nicoyans' roots to the indigenous Chorotega and their traditions have enabled them to remain relatively free of stress. Their traditional diet of fortified maize and beans may be the best nutritional combination for longevity the world has ever known.

Your Personal Blue Zone

Your Personal Blue Zone

Putting the Blue Zones Lessons to Work in Your Life

YOU'VE JUST READ STORIES ABOUT THE remarkable people of the world's Blue Zones. You've taken the time to get to know them and perhaps feel inspired by their experiences. Maybe you've noticed that the world's longevity all-stars not only live longer, they also tend to live better. They have strong connections with their family and friends. They're active. They wake up in the morning knowing that they have a purpose, and the world, in turn, reacts to them in a way that propels them along. An overwhelming majority of them still enjoy life. And there's not a grump in the bunch. But what does all this mean for you?

If you live the average American lifestyle, you may never reach your potential maximum lifespan. You might even fall short by as much as a decade. But what if you could follow a simple program that could help you feel younger, lose weight, maximize your mental sharpness, and keep your body working as long as possible? Indeed, what if you could get back that extra decade of healthy life that you may unknowingly be squandering?

This chapter presents the "Power Nine"—the lessons from the Blue Zones, a cross-cultural distillation of the world's best practices in health and longevity. While these practices are only associated with longevity and don't necessarily increase it, by adopting them you'll be adopting healthy habits that should stack the deck in your favor.

FIRST STEPS
Start down the road to longevity

To begin, go to the Blue Zones website at www.bluezones. com. There, you'll find the the Vitality Compass™, a tool that asks you 33 questions and, based on your answers, calculates 1) your potential life expectancy at your current age, 2) your healthy life expectancy—the number of good years you can expect to live, 3) the number of extra years you're likely to gain if you optimize your lifestyle, and 4) a customized list of suggestions to help you with that plan. Completing the Vitality Compass™ is the first step to figuring out where you are on your personal longevity journey.

The second step is to create a pro-longevity environment—your own personal Blue Zone—in your home. The goal will be to make positive behaviors convenient and, in some cases, unavoidable. Our strategies will put you in the way of pro-longevity practices so that if you make the effort now, you won't have to think about it later. We'll recommend nine deceptively simple but powerful things you can do today to create a lasting Blue Zone in your own life. They're designed to reinforce the lessons you've learned in this book without your having to keep track of anything.

The Power Nine covers the following life domains: What to *do* to optimize your lifestyle for a longer, healthier life; how to *think*; how to *eat*; and how to build *social relationships* that support your good habits. These lessons are patterned after the lifestyles of Blue Zones centenarians but modified to fit the Western lifestyle.

Research shows that if you dedicate yourself to a new practice for as little as five weeks, the practice is more likely to become a habit. (Another school of thought, called Relapse Prevention, suggests that for indulgent or addictive behaviors—overeating, gambling, drug use—the first three months of the initial change in behavior are crucial. If you make it past those first 12 weeks, your chances of relapse are greatly reduced.) So there's probably a time frame for behaviors to become habitual that ranges from 5 to 12 weeks.

Dr. Leslie Lytle of the University of Minnesota, a registered dietitian with a Ph.D. in health behavior, says that all of the habits we're recommending are relatively easy to adopt once you commit to making them work. She offers this advice to help you succeed: Pick the low-hanging fruit. All nine lessons offer the chance to gain more good years, so pick the ones that are easiest to do first. For example, try something you may have been successful at doing in the past. If you could do it then, focusing on it now might be easier. Another tip: Don't try more than three secrets at a time. If you work on all nine once, you'll be more likely to fail. Start with three that have the best chance of success, and then gradually add more as a pattern of success emerges. Lytle also recommends that you enlist a friend or family member in your program. If you hold each other accountable to a 12-week goal, you'll have a greater chance of succeeding. Remember to reward yourself, she says. Don't focus on your slip-ups; behavior change is hard at first. Focus on your small victories and celebrate them!

At first blush, this à la carte approach may seem too easy to work. But it reflects what current research tells us about effective long-term behavioral change. We're not proposing a timed program in this book because many people find it difficult to follow a highly prescriptive plan. Our lives are busy and often complicated, and a rigid plan with weekly goals may be too difficult to follow.

Early successes help keep us motivated, while early failure can make anything too difficult. If a nine-week program demands that you do something as simple as walk ten minutes a day during the first week, but you know you have trouble walking or you live in a neighborhood where you don't feel comfortable, you may not be able to succeed at that small change, and the whole plan may fall by the wayside. Even if we are committed to change, failing early can lead to discouragement and quitting.

Our approach allows you to pick and choose among the strategies that most appeal to you from the start. We emphasize changing your environment to help shape good habits. See where you can make small changes that will help create your Blue Zone. It's much easier to maintain healthy habits if your environment is set up for them.

LESSON ONE: MOVE NATURALLY

Be active without having to think about it

Longevity all-stars don't run marathons or compete in triathlons; they don't transform themselves into weekend warriors on Saturday morning. Instead, they engage in regular, low-intensity physical activity, often as part of a daily work routine. Male centenarians in Sardinia's Blue Zone worked most of their lives as shepherds, a profession that involved miles of hiking every day. Okinawans garden for hours each day, growing food for their tables. Adventists take nature walks. These are the kinds of activities longevity experts have in mind when they talk about exercising for the long haul. "The data suggests that a moderate level of exercise that is sustained is quite helpful," says Dr. Robert Kane.

An ideal routine, which you should discuss with your doctor, would include a combination of aerobic, balancing, and muscle-strengthening activities. Dr. Robert Butler recommends exercising the core muscle groups at least twice a week. Balance is also important since falls are a common cause of injuries and death among seniors (in the U.S.A., about one in three adults over age 65 falls each year). Just practicing standing on one foot during the day (maybe when brushing your teeth) would be a small step toward improving your balance.

Yoga, when done properly, will help increase balance. It also strengthens all muscle groups, increases flexibility, is good for your joints, and can lessen lower back pain. It may also provide the same kind of social support and spiritual centering that religions do.

In all longevity cultures, regular, low-intensity activity satisfies all of the above, while going easy on the knees and hips. Dr. Kane says the name of the game here is sustaining the effort. "You've got to be a miler, not a sprinter. You can't say, 'This year I'm really going to exercise like hell, and I can take it easy next year because I've done my exercise.'" The overall goal is to get into the habit of doing at least 30 minutes (ideally at least 60 minutes) of exercise at least five times a week. It doesn't have to be all at once, although that seems to be better.

LESSON ONE STRATEGIES

To get moving in your Blue Zone, try some of these tips.

Inconvenience yourself.

By making life a little tougher, you can easily add more activity to your days. Little things, like getting up to change the channel or taking the stairs, can add up to a more active lifestyle. Get rid of as many of the following as possible: TV remote control, garage door opener, electric can opener, electric blender, snow blower, and power lawn mower. Be ready to use as many of these as possible: bicycle, comfortable walking shoes, rake, broom, snow shovel.

Have fun. Keep moving.

Make a list of physical activities you enjoy. Rather than exercising for the sake of exercising, make your lifestyle active. Ride a bicycle instead of driving. Walk to the store. At work, take a walking break instead of a coffee and donut break. Build activity into your routine and lifestyle. Do what you enjoy. Forget the gym if you don't like it—you're not likely to go there if it's a chore. Don't force yourself to do things you dislike.

Walk.

This is the one activity that all successful centenarians did—and do—almost daily. It's free, easier on the joints than running, always accessible, invites company, and if you're walking briskly, may have the same cardiovascular benefits as running. After a hard day, a walk can relieve stress; after a meal, it can aid digestion.

Make a date.

Getting out and about can be more fun with other people. Make a list of people to walk with; combining walking and socializing may be the best strategy for setting yourself up for the habit. Knowing someone else is counting on you may motivate you to keep a walking date. A good place to start is to think, whose company do I enjoy? Who do I like to spend time with? Who has about the same level of physical ability?

Plant a garden.

Working in a garden requires frequent, low-intensity, full-range-of-motion activity. You dig to plant, bend to weed, and carry to harvest. Gardening can relieve stress. And you emerge from the season with fresh vegetables—a Blue Zones trifecta!

Enroll in a yoga class.

Be sure to practice it at least twice weekly.

LESSON TWO: HARA HACHI BU

Painlessly cut calories by 20 percent

If you're ever lucky enough to eat with Okinawan elders, you'll invariably hear them intone this Confucian-inspired adage before eating: hara hachi bu—a reminder to stop eating when their stomachs are 80 percent full. Even today, their average daily intake is only about 1,900 calories (Sardinians traditionally ate a similarly lean diet of about 2,000 calories a day).

Dr. Craig Willcox postulates that this simple but powerful practice may amount to a painless version of caloric restriction—a strategy that has been shown to prolong life in laboratory animals and has been associated with better heart health in humans. Some of the benefits of cutting calories may lie in reduced cellular damage from free radicals. But there's another happy by-product: You lose weight. Losing just 10 percent of one's body weight helps to lower blood pressure and cholesterol, which reduces the risk of heart disease. But how do we achieve this? Most of us don't live on a Japanese archipelago or surrounded by thousand-year-old cultural norms.

Our traditional solution for an expanding waistline is to start a diet. None of the centenarians we met were ever on a diet, and none of them were ever obese. "No diet yet studied works for most people," says University of Minnesota's Dr. Bob Jeffery. "You can get a diet to work for about 6 months, and then about 90 percent of all dieters run out of gas." Even the best programs are effective in the long run for only a small percentage of participants.

A secret to eating right for the long run is emulating the environment and habits of the world's longest-lived people. Dr. Brian Wansink, author of *Mindless Eating*, is conducting perhaps the most innovative research on what makes us eat the way we do. As Okinawan elders instinctively know, the amount of food we eat is less a function of feeling full and more a matter of what's around us. We overeat because of circumstances—friends, family, packages, plates, names, numbers, labels, lights, colors, candles, shapes, smells, distractions, cupboards, and containers.

In one of Wansink's experiments, he invited a group of people to view a tape and gave each one either a one-pound bag of M&Ms or a half-pound bag of M&Ms to eat as they pleased. After the video, he asked both groups to return their uneaten portions of candy. Those given a one-pound bag ate an average of 137 M&Ms, while those given a half-pound bag ate only 71. We typically consume more from big packages. Wansink's lab examined 47 products in this experiment, and the same thing happened over and over. He also found that the size of the plates and glasses we use has a profound impact on how much we consume. About three-quarters of what we eat is served on plates, bowls, or in glasses. Wansink's experiments showed that people drink 25 percent to 30 percent more if they drink from a short, wide glass rather than from a tall, narrow one, and 31 percent more if they eat from a 34-ounce bowl compared to a 17-ounce one.

While most Americans keep eating until their stomachs feel full, Okinawans stop as soon as they no longer feel hungry. "There's a significant calorie gap between when an American says, 'I'm full' and an Okinawan says, 'I'm no longer hungry,'" explains Wansink. "We gain weight insidiously, not stuffing ourselves, but eating a little bit too much each day—mindlessly."

Most of us have a caloric "set point" of sorts, a level of calories we can consume each day without gaining weight. We tend to gain weight by eating just slightly beyond the caloric set point week after week. For most of us, the solution is to eat enough so that we're no longer hungry, but not so much that we're full. Wansink asserts that we can

Okinawans, like this 88-year-old fisherman, often remind themselves not to overeat.

eat about 20 percent more or 20 percent less without really being aware of it. And that 20 percent swing is the difference between losing weight and gaining it.

Food volume is only one part of the equation. The other is calories. A typical fast-food meal of a large hamburger, large fries, and a large soft drink contains nearly 1,500 calories. Craig and Bradley Willcox estimate that the average Okinawan meal has one-fifth the caloric density. In other words, a hamburger with fries and a heaping plate of Okinawan stir-fried tofu and greens may have the same volume, but the Okinawan meal will have one-fifth the calories.

Most people are terrible at estimating how many calories they've eaten today, yesterday, or last week, Wansink cautions. Caloric content of food is typically underestimated by about 20 percent. The larger the meal, the worse the underestimation. In one of his studies, Wansink had people estimate how many calories were in a meal. It actually contained about 1,800 calories, but the average person's guess was around 1,000 calories. The trick to maintaining a healthy weight is to eat foods with low caloric density. If we look at our food and think it will fill us, it probably will. As Wansink writes, volume trumps calories.

LESSON TWO STRATEGIES

To help follow the 80 percent rule in your Blue Zone, try the following tips.

Serve and store.
People who serve themselves at the counter, then put the food away before taking their plate to the table, eat about 14 percent less than when they take smaller amounts and go back for seconds and thirds. Learn to recognize when you have enough on your plate to fill your stomach 80 percent.

Make food look bigger.
People who eat a quarter-pound hamburger that has been made to look like a half-pounder with lettuce, tomatoes, and onions feel equally full after eating. Students who drank a smoothie whipped to twice the volume with the same calories ate less for lunch 30 minutes later, and reported feeling fuller.

Use small vessels.
Drop your dinner plates and big glasses off at a charity, and buy smaller plates and tall, narrow glasses. You're likely to eat significantly less without even thinking about it.

Make snacking a hassle.

Avoid tempting foods. Put candy bowls, cookie jars, and other temptations out of sight. Hide them in the cupboard or pantry. Wrap tempting leftovers in an opaque container.

Buy smaller packages.

When given large packages of spaghetti, sauce, and meat, Wansink's subjects consumed 23 percent more (about 150 calories) than when they were given medium packages.

Give yourself a daily reminder.

The bathroom scale can be a simple yet powerful reminder not to overeat. Put the scale in your way so you can't avoid a daily weigh-in. In fact, weighing yourself is one of the most surefire ways to reduce your weight and keep it off in the long run. One study that followed 3,026 women who were trying to lose weight and keep it off found that after 2 years, women who weighed themselves daily lost an average of 12 pounds. Women who never weighed in actually gained an average of 5 pounds. In other words, at the end of 2 years, women who weighed themselves every day were (on average) about 17 pounds lighter than the women who never weighed themselves.

Eat more slowly.

Eating faster usually results in eating more. Slowing down allows time to sense and react to cues telling us we're no longer hungry.

Focus on food.

A guaranteed way to eat mindlessly is to do so while watching your favorite show on TV or while emailing a friend at the computer. If you're going to eat, just eat. You'll eat more slowly, consume less, and savor your food more.

Have a seat.

Many of us eat on the run, in the car, standing in front of the refrigerator, or while walking to our next meeting. This means we don't notice what we're eating or how fast we're eating it. Making a habit of eating only while sitting down—eating purposefully—better enables us to appreciate the tastes and textures of our food. We'll eat more slowly and feel more nourished when we're through.

Eat early.
In the Blue Zones, the biggest meal of the day is typically eaten during the first half of the day. Nicoyans, Okinawans, and Sardinians eat their biggest meal at midday, while Adventists consume many of their calories for breakfast. All Blue Zone residents eat their smallest meal of the day in late afternoon or early evening. Some Adventists believe that if you eat a big breakfast with the right ingredients (whole grains, fruits, milk, nut butter), you'll fuel your body for most of the day and have fewer cravings for sugary or fatty foods.

LESSON THREE: PLANT SLANT
Avoid meat and processed foods

Most centenarians in Nicoya, Sardinia, and Okinawa never had the chance to develop the habit of eating processed foods, soda pop, or salty snacks. For much of their lives, they ate small portions of unprocessed foods. They avoided meat— or more accurately, didn't have access to it—except on rare occasions. Traditional Sardinians, Nicoyans, and Okinawans ate what they produced in their gardens, supplemented by staples: durum wheat (Sardinia), sweet potato (Okinawa), or maize (Nicoya). Strict Adventists avoid meat entirely. They take their dietary cues directly from Genesis 1:29 . . .

> Then God said, "Behold, I have given you every plant yielding seed that is on the surface of all the earth, and every tree which has fruit yielding seed; it shall be food for you."

Indeed, scientists analyzed six different studies of thousands of vegetarians and found that those that restrict meat are associated with living longer. Some people worry that a plant-based diet may not provide enough protein and iron. The fact is, says Dr. Leslie Lytle, that those of us over 19 years of age need only 0.8 grams of protein for every kilogram (2.2 pounds) of our weight, which for most of us would amount to only 50 to 80 grams (1.8 to 2.8 ounces) of protein daily.

"Our bodies can't store protein," she says. "Extra protein gets converted to calories, and if not needed for activity or to maintain our bodies, it eventually becomes fat. While we don't need a lot of protein in our diets, we should eat some protein at every meal. Protein helps us feel fuller, and helps us avoid the peaks and valleys in our blood sugar levels that make us feel hungry." Similarly, most of us (except pre-menopausal women) can get plenty of iron from fortified grains. Too much iron can actually be bad for us because it may play a role in generating oxygen-free radicals.

Beans, whole grains, and garden vegetables are the cornerstones of all these longevity diets. Sardinian shepherds take semolina flatbread into the pastures with them. Nicoyans eat corn tortillas at every meal. And whole grain is part of the Adventist diet. Whole grains deliver fiber, antioxidants, potential anti-cancer agents (insoluble fiber), cholesterol reducers, and clot blockers, plus essential minerals. Beans (legumes) also provide a cornerstone to Blue Zone meals. Diets rich in legumes are associated with fewer heart attacks and less colon cancer. Legumes are a good dietary source of healthy flavonoids and fiber (which can reduce the risk of heart attack) and are also an excellent nonanimal source of protein.

Tofu (bean curd), a daily feature in the Okinawan diet, has been compared to bread in France or potatoes in Eastern Europe, the difference being that while one cannot live by bread or potatoes alone, tofu is an almost uniquely perfect food: low in calories, high in protein, rich in minerals, devoid of cholesterol, eco-friendly, and complete in the amino acids necessary for human sustenance. An excellent source of protein without the side effects of meat, tofu contains a compound, phytoestrogen, which may provide heart-protective properties to women. In addition, phytoestrogen seems to modestly lower cholesterol and promote healthy blood vessels.

All of which is not to say that people who live a long time don't eat meat: Festive meals in Sardinia include lots of meat. Okinawans slaughter pigs during lunar New Year festivals. And Nicoyans raise family pigs as well. But meat is typically eaten only a few times a month. And most warnings concern red meat or processed meat like bacon. Both Dr. Robert Kane and Dr. Robert Butler say that when establishing a diet, it's important to balance calories between complex carbohydrates, fats, and proteins, minimizing trans fats, saturated fats, and salt.

"The key is taking in what you need and avoiding the extremes," explains Dr. Kane. "You could go to a nonanimal diet, but an awful lot of vegetarians take in more cholesterol through cheeses and milk and various products than they ever would take in eating meat at a modest level." Oddly, pork was common to three of the four Blue Zones diets. But it was not eaten regularly. Nuts are perhaps the most impressive of all "longevity foods." Recent findings from a large study of Seventh-day Adventists show that those who ate nuts at least five times a week had a rate of heart disease that was half that of those who rarely ate nuts. A health claim about nuts is among the first qualified claims permitted by the Food and Drug Administration. In 2003, the FDA allowed a "qualified health claim" that read: "Scientific evidence suggests but does not prove that eating 1.5 ounces per day of most nuts as part of a diet low in saturated fat and cholesterol may reduce the risk of heart disease."

Studies have indicated that nuts may help protect the heart by reducing total blood cholesterol levels. In a large, ongoing population study from Harvard University's School of Public Health, people who often ate nuts had lower risks of coronary heart disease than those who rarely or never ate nuts. The Adventist Health Study (AHS) showed that the person who ate nuts at least five times per week, two ounces per serving, lived on average about two years longer than those who didn't eat nuts.

"We don't know if there's something magical in nuts or if it's that people who are eating nuts are not eating junk food. But the positive effect is clear from the data," says Dr. Gary Fraser, director of the AHS study. One explanation might be that nuts are rich in monounsaturated fat and soluble fiber, both of which tend to lower LDL cholesterol, he says. They are also relatively good sources of vitamin E and other possibly heart-protective nutrients. The best nuts are almonds, peanuts, pecans, pistachios, hazelnuts, walnuts, and some pine nuts. Brazil nuts, cashews, and macadamias have a little more saturated fat and are less desirable. But all nuts are good.

LESSON THREE STRATEGIES

Try these tips to incorporate more plants in your diet.

Eat four to six vegetable servings daily.
Blue Zone diets always include at least two vegetables at each meal.

Limit intake of meat.
The centenarians in the Blue Zones consume limited quantities of meat. To emulate their diet, try to limit serving meat to twice weekly, and serve no portion larger than a deck of cards.

Showcase fruits and vegetables.
Put a beautiful fruit bowl in the middle of your kitchen table. At the bottom of it, leave a note that reads "Fill Me." Instead of hiding vegetables in the refrigerator compartment that says "Produce," put them front and center on shelves where you can see them.

Lead with beans.
Beans are a cornerstone of each of the Blue Zone diets. Make beans—or tofu—the centerpiece of lunches and dinners.

Eat nuts every day.
The Adventist Health Study shows that it doesn't matter what kind of nuts you eat to help extend your life expectancy. Caution: A 1-ounce serving of nuts typically ranges from 165 to 200 calories, so 2 ounces could be almost 400 calories.

Stock up.
Keep nuts available in your home, ideally in packages of two ounces or less. (You might want to store them in the refrigerator to keep oils fresh.) Or keep a can of nuts in the office for an afternoon snack, which may help avoid snacking right before dinner.

LESSON FOUR: GRAPES OF LIFE
Drink red wine (in moderation)

Epidemiological studies seem to show that people who have a daily drink per day of beer, wine, or spirits may accrue

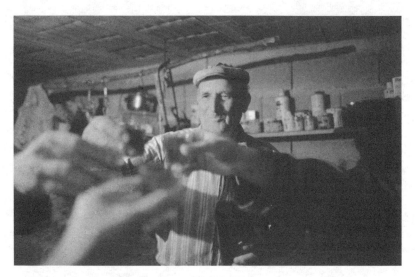

Sardinians share a toast with glasses of Cannonau, a locally produced red wine that has higher concentrations of polyphenols.

some health benefits. But the secrets of the Blue Zones suggest that consistency and moderation are key. In Okinawa, it's a daily glass of sake with friends. In Sardinia, it's a glass of dark red wine with each meal and whenever friends meet.

A daily drink or two has been associated with lower rates of heart disease, but alcohol use has also been shown to increase the risk of breast cancer, for instance. It does appear to reduce stress and the damaging effects of chronic inflammation. What's more, having a glass of wine with a meal creates an "event," and thus makes it more likely you'll eat more slowly.

Red wine offers an extra bonus in that it contains artery-scrubbing polyphenols that may help fight arteriosclerosis. For an extra antioxidant bump, choose a Sardinian Cannonau. Beware, however, of the toxic effects of alcohol on the liver, brain, and other organs—along with increased risk of accidents—when daily consumption exceeds a glass or two. In that case, the risks of drinking can outweigh any health benefits. A friend recently asked me if he could save up all week and drink 14 glasses of wine on Saturday night. The answer is no.

LESSON FOUR STRATEGIES

Introduce a glass of wine into a daily routine.

Buy a case of high-quality red wine.
The Sardinians quaff Cannonau in their Blue Zone. This variety might be hard to find in parts of the United States, but any dark red wine should do.

Treat yourself to a "Happy Hour."
Set up yours to include a glass of wine, nuts as an appetizer, and a gathering of friends or time with a spouse.

Take it easy.
A serving or two per day of red wine is the most you need to drink to take advantage of its health benefits. Overdoing it negates any benefits you might enjoy, so drink in moderation.

LESSON FIVE: PURPOSE NOW
Take time to see the big picture

Okinawans call it *ikigai*, and Nicoyans call it *plan de vida*, but in both cultures the phrase essentially translates to "why I wake up in the morning." The strong sense of purpose possessed by older Okinawans may act as a buffer against stress and help reduce their chances of suffering from Alzheimer's disease, arthritis, and stroke.

Dr. Robert Butler and collaborators led an NIH-funded study that looked at the correlation between having a sense of purpose and longevity. His 11-year study followed highly functioning people between the ages of 65 and 92 and found that individuals who expressed a clear goal in life—something to get up for in the morning, something that made a difference—lived longer and were sharper than those who did not. It was also reported that immediately following December 31, 1999, demographers saw a spike in deaths among elders. These older people, in other words, may have willed themselves to stay alive into the new millennium.

A sense of purpose may come from something as simple as seeing that children or grandchildren grow up well. Purpose can come from a job or a hobby, especially if you can immerse yourself completely in it. Claremont University psychologist Dr. Mihály Csíkszentmihályi best describes this feeling in his book, *Flow: The Psychology of Optimal Experience*. He defines flow as a Zen-like state of total oneness with the activity at hand in which you feel fully immersed in what you're doing. It's characterized by a sense of freedom, enjoyment, fulfillment, and skill, and while you're in it, temporal concerns (time, food, ego/self, etc.) are typically ignored. If you can identify the activity that gives you this sense of flow and make it the focus of your job or hobby, it can also become your sense of purpose.

A new activity can give you purpose as well. Learning a musical instrument or a new language gives you a double bonus, since both have been shown to help keep your brain sharper longer. "Exercising your brain is important," says Dr. Thomas Perls of Boston University Medical School.

"Doing things that are novel and complex. Once you get good at them, and they are no longer novel, then you move on to something else. So you're kind of doing strength training for the brain, and that has been shown to decrease your rate of memory loss, and maybe even decrease the rate at which one might develop Alzheimer's disease."

LESSON FIVE STRATEGIES

To realize your purpose, try the following tips.

Craft a personal mission statement.

If you don't have a sense of purpose, how do you find it? Articulating your personal mission statement can be a good start. Begin by answering this question in a single, memorable sentence: Why do you get up in the morning? Consider what you're passionate about, how you enjoy using your talents, and what is truly important to you.

Find a partner.

Find someone to whom you can communicate your life purpose, along with a plan for realizing it. It can be a friend, a family member, a spouse, a colleague—as long as it's someone who can help you honestly assess your plan and your successes.

Learn something new.

Take up a musical instrument or learn a new language. Both activities are among the most powerful things you can do to preserve your mental sharpness.

LESSON SIX: DOWN SHIFT

Take time to relieve stress

Sardinians pour into the streets at 5 p.m., while Nicoyans take a break every afternoon to rest and socialize with friends. Remember Ushi and her moai? They gather every evening before supper to socialize. People who've made it to 100 seem to exude a sense of sublime serenity. Part of it is that their bodies naturally slow down as they have aged, but they're also wise enough to know that many of life's most precious moments pass us by if we're lurching blindly toward some goal. I remember watching Gozei Shinzato pause to watch a brilliant thunderstorm as she washed her breakfast dishes in Okinawa, and Sardinian shepherd Tonino Tola stop to take a long look over the emerald green plateaus below. He'd seen that same sweeping view for almost 80 years, yet still took time every day to appreciate it.

For Adventists, the Saturday Sabbath means many things. One is that the Sabbath can be a powerful stress reliever. From sunset Friday to sunset Saturday, Adventists create a "sanctuary in time" during which they focus on God, their families, and nature. They don't work. Kids don't play organized sports or do homework. Instead families do things together, such as hiking, that bring them together and make them feel closer to God. For the Adventists, it's a time to put the rest of the week in perspective and to lessen the din and confusion of everyday life.

The result seems to be a greater sense of well-being. But how does slowing down help you live longer? The answer may have something to do with chronic inflammation. Inflammation is the body's reaction to stress. That stress can come in the form of an injury, an infection, or anxiety. Small amounts of stress can be good—for fighting off disease, helping us heal, or preparing us for a traumatic event. But when we chronically trigger inflammation, our bodies can turn on themselves. Italian endocrinologist Dr. Claudio

Franceschi has developed a widely accepted theory on the relationship between chronic inflammation and aging. Over time, he believes, the negative effects of inflammation build up to create conditions in the body that may promote age-related diseases such as Alzheimer's disease, atherosclerosis, diabetes, and cardiovascular disease. Slowing life's pace may help keep the chronic inflammation in check, and theoretically, the related disease at bay.

Apart from such health benefits, this Blue Zone lesson does much to add richness to life. Slowing down ties together so many of the other lessons—eating right, appreciating friends, finding time for spirituality, making family a priority, creating things that bring purpose. I remember sitting one cloudy afternoon with Raffaella Monne in the Sardinian village of Arzana. Outliving most of her children, this woman had enjoyed a full life; at age 107, she stayed in her home for much of the day, venturing out to a nearby plaza some afternoons. Though she could only muster a whisper, her loving, serene demeanor attracted people. Children would make a point of visiting her on their way home from school.

The day we met, Raffaella sat peeling an apple as I peppered her with dozens of questions about her diet, level of physical exercise, the relationship to her family, and so forth. Her answers were laconic and unrevealing. Finally, exasperated, I asked her if after 107 years she had any advice for younger people. She looked up at me, eyes flashing. "Yes," she shot back. "Life is short. Don't run so fast you miss it."

Once again, though, we're confronted by an environmental catch-22. In the Western world, accomplishment, status, and material wealth are highly revered and require most of our time. Americans employed full time work on average 43 hours a week and take the shortest paid vacations in the industrialized world. Then when they do take time off, according to one source, 20 percent of them stay in touch with the office. We generally hold working and being productive in high regard; being busy often wins us esteem. Few cultural institutions exist to encourage us to slow down, unwind, and de-stress.

Finding time for our spiritual side can create a space to slow down, and practices like yoga and meditation can also give the mind a respite. Steve Hagen, who was ordained in the Soto Zen school of Buddhism and who wrote *Buddhism, Plain and Simple*, regards meditation as a cornerstone of slowing down. "Meditation provides us a mechanism to step out of the self-focus and find true freedom."

Regular meditation can allow us to slow down our minds, ridding them of the incessant chatter in our heads. It focuses concentration and allows us to see the world as it really is, instead of how we imagine it to be. It sets up our day and helps us to realize that rushing, worrying, and the urgency we give to so many things in our lives really aren't so important. With that realization, all other strategies for slowing down come much easier.

LESSON SIX STRATEGIES

Use these tips to find a quiet space to slow down in your Blue Zone.

Reduce the noise.
Minimizing time spent with television, radio, and the Internet can help reduce the amount of aural clutter in your life. Rid your home of as many TVs and radios as possible, or limit them to just one

room. Most electronic entertainment just feeds mind chatter and works counter to the notion of slowing down.

Be early.
Plan to arrive 15 minutes early to every appointment. This one practice minimizes the stress that arises from traffic, getting lost, or underestimating travel time. It allows you to slow down and focus before a meeting or event.

Meditate.
Create a space in your home that is quiet, not too hot and not too cold, not too dark and not too light. Furnish the space with a meditation cushion or chair. Establish a regular meditation schedule, and try to meditate every day no matter what (but also not stressing out if you fail to do so). Start with 10 minutes a day, and try to work up to 30 minutes a day. Try to meditate with others occasionally.

LESSON SEVEN: BELONG
Participate in a spiritual community

Healthy centenarians everywhere have faith. The Sardinians and Nicoyans are mostly Catholic. Okinawans have a blended religion that stresses ancestor worship. Loma Linda centenarians are Seventh-day Adventists. All belong to strong religious communities. The simple act of worship is one of those subtly powerful habits that seems to improve your chances of having more good years. It doesn't matter if you are Muslim, Christian, Jewish, Buddhist, or Hindu.

Studies have shown that attending religious services— even as infrequently as once a month—may make a difference in how long a person lives. A recent study in the *Journal of Heath and Social Behavior* followed 3,617 people for seven and half years and found that those who attended religious services at least once a month reduced their risk of death by about a third. As a group, the attendees had a longer life expectancy, with an impact about as great as that of moderate physical activity.

The NIH-funded Adventist Health Study had similar findings. It followed more than 34,000 people over a period of 12 years, and found that those who went to church services frequently were 20 percent less likely to die at any age. It appears that people who pay attention to their spiritual side have lower rates of cardiovascular disease, depression, stress, and suicide, and their immune systems seem to work better.

Put generally, the faithful are healthier and happier. What's the rationale? In his book *Diet, Life Expectancy and Chronic Disease: Studies of Seventh-day Adventists and Other Vegetarians,* Dr. Gary Fraser cited several reasons to support this claim. People who attend church are less likely to engage in harmful behaviors and more likely to take on healthful behaviors. They were physically more active, less likely to smoke, do drugs, or drink and drive. People who attend church have a forced schedule of self-reflection, decompression, and stress-relieving meditation, either through regular prayer or from sitting quietly during religious services.

Belonging to a religious community can foster larger and denser social networks. People who attend services may have higher self-esteem and self worth because religion encourages positive expectations, which in turn can actually improve health. When individuals undertake and successfully act in ways prescribed by their roles, their self-concept and sense of well-being are reinforced. To a certain extent, adherence to a religion allows them to relinquish the stresses of everyday

Seventh-day Adventists worship at the University Church in Loma Linda, California.

life to a higher power. A code of behavior is clearly laid out before them. If you follow it, you have the peace of mind that you're engaging in "right living." If today goes well, perhaps you deserve it. If today goes poorly, it's out of your hands.

LESSON SEVEN STRATEGIES

To strengthen the spiritual dimension of your Blue Zone, try these tactics.

Be more involved.

If you already belong to a religious community, take a more active role in the organization. The longevity-enhancing effect may be a function of how you attend rather than the fact that you just attend. Getting involved in activities like singing in the choir or volunteering might enhance well-being and possibly reduce mortality.

Explore a new tradition.

If you don't have a particular religious faith, commit to trying a new faith community. If you don't subscribe to any specific denomination, or if you haven't found a positive religious experience, you may want to explore a belief that is not based on strict dogma. Unitarian Universalism, for example, is open to anyone who believes in the inherent worth and dignity of every person and in the acceptance and encouragement of each individual's own spiritual journey. Buddhism is another tradition to consider if you are looking for a religious community. There is also the American Ethical Union, which describes itself as a "humanistic religious and educational movement." The Union was inspired by the ideal that the supreme aim in life is to create a more humane society. Members join together in ethical societies to assist one another in developing moral ideas.

Just go.

Schedule an hour a week for the next eight weeks to attend religious services. Don't think about it. Just go, and do so with an open mind. Studies show that people who get involved with the service (singing hymns, participating in prayers or liturgy, volunteering) may find their well-being enhanced.

LESSON EIGHT: LOVED ONES FIRST

Make family a priority

The most successful centenarians we met in the Blue Zones put their families first. They tended to marry, have children, and build their lives around that core. Their lives were imbued with familial duty, ritual, and a certain emphasis on togetherness. This finding was especially true in Sardinia, where residents still possess a zeal for family. I asked the owner of one Sardinian vineyard who was caring for his infirm mother if it wouldn't be easier to put her in a home. He wagged his finger at me, "I wouldn't even think of such a thing. It would dishonor my family."

Tonino Tola, the Sardinian shepherd, had a great passion for work, but told us, "Everything I do is for my family." In Nicoya, families tend to live in clusters. All 99 inhabitants of one village were descended from the same 85-year-old man. They still meet for meals at a family-owned mountaintop restaurant, and the patriarch's grandchildren still visit him daily to help him tidy up or just to play a game of checkers.

The Okinawan sense of family even transcends this life. Okinawans over 70 still begin the day by honoring their ancestors' memories. Gravesites are often furnished with picnic tables so that family members can celebrate Sunday meals with deceased relatives.

How does this contribute to longevity? By the time centenarians become centenarians, their lifelong devotion has produced returns: their children reciprocate their love and care. Their children check up on their parents, and in three of the four Blue Zones, the younger generation welcomes the older generation into their homes. Studies have shown that elders who live with their children are less susceptible to disease, eat healthier diets, have lower levels of stress, and have a much lower incidence of serious accidents. The

MacArthur Study of Successful Aging, which followed 1,189 people between the ages of 70 and 79 for more than 7 years, showed that elders who lived with their families had much sharper mental and social skills.

"Families represent the highest degree of social network," says Dr. Butler. "Parents that give you a sense of reality, of how to behave healthwise, offer a sense of goals and purpose, and then if you become ill, or problems emerge, that basic support of family becomes incredibly important." Much of life is making investments, he says. You make investments when you go to school and you get educated in a particular field. Investing in our children when they are young helps assure they'll invest in us when we're old. The payoff? Seniors who live with their families stay sharper longer than those who live alone or in a nursing home.

America is trending in the opposite direction. In many busy families with working parents and active kids, family time can become rare as everyone's schedules become more and more packed with things to do. Shared meals and activities can drop off the daily routine, making time together difficult to come by.

How do you buck the trend? Gail Hartman, a licensed psychologist, believes the key is for all generations of a family to make of point of spending time together. "Successful families make a point of eating at least one meal a day together, taking annual vacations, and spending family time. Everything does not need to stop. Kids can be doing homework, and parents can be preparing dinner, but the point is that there is a 'we-ness' to the family."

LESSON EIGHT STRATEGIES

These tips can help you create your family's Blue Zone.

Get closer.
Consider living in a smaller house to create an environment of togetherness. A large, spread-out house makes it easier for family members to segment themselves from the group. It's easier for families to bond and spend time together in a smaller home. If you live in a large home, establish one room where family members gather daily.

Establish rituals.
Children thrive on rituals; they enjoy repetition. Make one family meal a day sacred. Establish a tradition for a family vacation. Have dinner with Grandma every Tuesday night. Make a point to purposefully celebrate holidays.

Create a family shrine.
In Okinawan homes, the ancestor shrine is always displayed in the best room in the house. It showcases pictures of deceased loved ones and their important possessions. It serves as a constant reminder that we're not islands in times but connected to something bigger. We can pick a wall to display pictures of our parents and children, or take annual family pictures and display them in progression.

Put family first.
Invest time and energy in your children, your spouse, and your parents. Play with your children, nurture your marriage, and honor your parents.

LESSON NINE: RIGHT TRIBE

Be surrounded by those who share Blue Zone values

This is perhaps the most powerful thing you can do to change your lifestyle for the better. For residents of the Blue Zones it comes naturally. Seventh-day Adventists make a point of associating with one another (a practice reinforced by their religious practices and observation of the Sabbath on Saturdays). Sardinians have been isolated geographically in the Nuoro highlands for 2,000 years. As a result, members of these longevity cultures work and socialize with one another, and this reinforces the prescribed behaviors of their cultures. It's much easier to adopt good habits when everyone around you is already practicing them.

Social connectedness is ingrained into the world's Blue Zones. Okinawans have moais, groups of people who stick together their whole lives. Originally created out of financial necessity, moais have endured as mutual support networks. Similarly, Sardinians finish their day in the local bar where they meet with friends. The annual grape harvest and village festivals require the whole community to pitch in.

Professor Lisa Berkman of Harvard University has investigated social connectedness and longevity. In one study, she looked at the impact of marital status, ties with friends and relatives, club membership, and level of volunteerism on how well older people aged. Over a nine-year period, she found that those with the most social connectedness lived longer. Higher social connectedness led to greater longevity. Those with the least social connectedness were between two and three times more likely to die during the nine-year period of the study than those with the most social connectedness. The type of social connectedness was not important in relation to longevity—as long as there was connection. Even a lack of a spouse or significant other could be compensated for by other forms of connection.

A recent article in the *New England Journal of Medicine* showed just how powerful an immediate social network can be. Looking at a community of 12,067 people over a period of 32 years, researchers found that subjects were more likely to become obese when their friends became obese. In the case of close mutual friends, if one became obese, the odds of the other becoming obese nearly tripled. It seemed the same effect occurred with weight loss.

"I think a superior social support network is one of the reasons that women live longer than men," says Dr. Robert Butler. "They have better and stronger systems of support than men, they're much more engaged with and helpful to each other, more willing and able to express feelings, including grief and anger, and other aspects of intimacy."

LESSON NINE STRATEGIES

Try these tips to build up the inner circle of your Blue Zone.

Identify your inner circle.

Know the people who reinforce the right habits, people who understand or live by Blue Zone secrets. Go through your address book or your contact list of friends. Think about which ones support healthy habits and challenge you mentally, and which ones you can truly rely on in case of need. Put a big "BZ" by their names. Ideally, family members are the first names on that list.

Be likable.

Of the centenarians interviewed, there wasn't a grump in the bunch. Dr. Nobuyoshi Hirose, one of Japan's preeminent longevity experts, had a similar observation. Some people are born popular, and people are naturally drawn to them. Likable old people are more likely to have a social network, frequent visitors, and de facto caregivers. They seem to experience less stress and live purposeful lives.

Create time together.

Spend at least 30 minutes a day with members of your inner circle. Establish a regular time to meet or share a meal together. Take a daily walk. Building a strong friendship requires some effort, but it is an investment that can pay back handsomely in added years.

Fumiyasu Yamakawa, 84, finds his ikigai, or purpose, in his daily exercises.

THE CHOICE IS UP TO US

I'd like to end this book with a tribute. The very first centenarian I met was a 112-year-old Seventh-day Adventist named Lydia Newton. At the time, she was one of the 35 oldest people on the planet. I traveled to the outskirts of Sedona, Arizona, to talk to her in the triple-wide mobile home that she shared with her 91-year-old daughter, Margarite. When I walked into the living room, Lydia was wearing a dress with embroidered lilies that she had sewn herself 70 years earlier.

"Now, what kind of dirt are they spreading about me?" she said by way of greeting and chuckled. I spent two days with her, asking her questions about her lifestyle (she started her day with buttermilk, oatmeal, and quiet prayer), listening to stories (she recounted a 107-year-old memory of a bull goring her father, including a description of the "crimson" blood staining his cotton shirt), and watching her as she carried out her everyday life. She and Margarite were best friends; they could sit and talk and find each other's company completely entertaining.

When I left, I asked if I could hug her. She said of course. As I bent down, nearing her face, I could see her veins through parchment-paper-thin skin. I embraced her, feeling the birdlike bones of her back, and felt the universal warmth of life within her embrace. It felt like hugging a delicate child.

"The way you live," I told her, "I'd be surprised if you don't make the Guinness World Records for the oldest person who ever lived."

"Oh, go on," she said with a wave of supreme indifference.

A few months into writing this book, I got a call from Margarite. Her mother had died. Lydia had peacefully dozed off and never woke up. She was the first of three centenarian friends I met while writing this book who died before I finished it. None of them wanted to die, but they didn't fear it either. They'd all mastered the art of life and accepted the inevitability of its end.

To me, they offered a lesson about decline. I know that our bones will soften and our arteries will harden. Our hearing will dull and our vision will fade. We'll slow down. And, finally, our bodies will fail altogether, and we'll die. How this decline unfolds is up to us. The calculus of aging offers us two options: We can live a shorter life with more years of disability, or we can live the longest possible life with the fewest bad years.

As my centenarian friends showed me, the choice is largely up to us.

Bibliography

BOOKS

Corder, Roger. *The Red Wine Diet*. Avery, 2007.

Csíkszentmihályi, Mihály. *Finding Flow: The Psychology of Engagement with Everyday Life*. Basic Books, 1997.

Fraser, Gary E. *Diet, Life Expectancy, and Chronic Disease: Studies of Seventh-day Adventists and Other Vegetarians*. Oxford University Press, 2003.

Koenig, Harold. *Is Religion Good for Your Health? The Effects of Religion on Physical and Mental Health*. The Haworth Pastoral Press, 1997.

Lawrence, D. H. *Sea and Sardinia*. New edition. Penguin, 1999.

Pratt, Steven, M.D. and Kathy Matthews. *SuperFoods Rx: Fourteen Foods That Will Change Your Life*. Harper, 2004.

Robbins, John. *Healthy at 100*. Random House, 2006.

Wagner, Philip. *Nicoya: A Cultural Geography*. University of California Press, 1958.

Wansink, Brian. *Mindless Eating: Why We Eat More Than We Think*. Bantam Books, 2006.

Weil, Andrew. *Healthy Aging: A Lifelong Guide to Your Physical and Spiritual Well-Being*. Knopf, 2005.

Willcox, Bradley J., D. Craig Willcox, and Makoto Suzuki. *The Okinawa Program: How the World's Longest-Lived People Achieve Everlasting Health—and How You Can Too*. Clarkson Potter, 2001.

Willcox, Bradley J., D. Craig Willcox, and Makoto Suzuki. *The Okinawa Diet Plan: Get Leaner, Live Longer, and Never Feel Hungry*. Three Rivers Press, 2005.

ARTICLES

Buettner, Dan. "The Secrets of Living Longer." *National Geographic* (November 2005), 2-27.

Butler, Robert N. "The inequality of longevity: Life expectancy gap widens between industrialized world and developing nations." *Geriatrics* (December 1999).

"Can We Live to 150?" *Popular Science* (November 1993), 77-82.

"Cancer Prevention and Early Detection: Facts and Figures," American Cancer Society (2007). Available online at http://www.cancer.org/docroot/STT/content/STT_1x_Cancer_Facts__Figures_2007.asp

Christakis, N. A., and J. H. Fowler. "The spread of obesity in a large social network over 32 years." *The New England Journal of Medicine* (July 26, 2007), 370-79.

Clarkson, Thomas B. "Soy, Soy Phytoestrogens, and Cardiovascular Disease." Supplement: Fourth International Symposium on the Role of Soy in Preventing and Treating Chronic Disease." *Journal of Nutrition* (March 2002), 566S-569S.

Corder, Roger, W. Mullen, N. Q. Khan, S. C. Marks, E. G. Wood, M. J. Carrier, and A. Crozier. "Oenology: Red Wine Procyanidins and Vascular Health." *Nature* (November 30, 2006), 566.

Corliss, Richard, and Michael D. Lemonick. "How To Live To Be 100" *Time* (August 30, 2004), 40.

Craun, G. F., and R. L. Calderon. "How to interpret epidemiological associations." in *Rolling Revision of the WHO Guidelines for Drinking-Water Quality* (draft for review and comments). World Health Organization (August 2004). Available online at http://www.who.int/water_sanitation_ health/dwq/nutepiassociations.pdf

Fontana, Luigi, Dennis T. Villareal, Edward P. Weiss, Susan B. Racette, et al. "Calorie restriction or exercise: effects on coronary heart disease risk factors. A randomized, controlled trial." *American Journal of Physiology* (July 2007), E197-E202.

Foreyt, John P. "Diet, Behavior Modification, and Obesity: Nine Questions Most Often Asked by Physicians." *Consultant* (June 1990), 53-6.

Fraser, Gary E., and David J. Shavlik. "Ten Years of Life: Is It a Matter of Choice?" *Archives of Internal Medicine* (July 9, 2001), 1645-52. Available online at http://archinte.ama-assn.org/cgi/content/full/161/13/ 1645?lookupType=volpage&vol=161&fp=1645&vicw=full

Haber, Carole. "Life Extension and History: The Continual Search for the Fountain of Youth." *The Journals of Gerontology: Series A* (June 2004), B515-22.

He, Wan, Manisha Segupta, Victoria A. Velkoff, and Kimberly A. DeBarros. "65 + in the United States: 2005." U.S. Census Bureau (December 2005). Available online at http://www.census.gov/prod/2006pubs/p23-209.pdf

Heilbronn, Leonie K., Lilian de Jonge, Medlyn I. Frisard, James P. DeLany, et al. "Effect of 6-Month Calorie Restriction on Biomarkers of Longevity, Metabolic Adaptation, and Oxidative Stress in Overweight Individuals: A Randomized Controlled Trial." *JAMA* (April 5, 2006), 1539-48.

Herskind, Anne Maria, Matthew McGue, Niels V. Holm, Thorkild I. A. Sørensen, Bent Harvald, and James W. Vaupel. "The heritability of human longevity: A population-based study of 2,872 Danish twin pairs born 1870–1900." *Human Genetics* (March 1996) 319-23.

Hilton, Lisette. "Tobacco isolated as cause of skin aging." *Dermatology Times* (September 2002), 20: 23.

Hjelmborg, Jacob v.B., Ivan Iachine, Axel Skytthe, James W. Vaupel, Matt McGue, Markku Koskenvuo, Jaakko Kaprio, Nancy L. Pedersen, and Kaare Christensen. "Genetic influence on human lifespan and longevity." *Human Genetics* (April 2006), 312-21.

Kemper, Peter, Harriet L. Komisar, and Lisa Alecxih. "Long-Term Care Over an Uncertain Future: What Can Current Retirees Expect?" *Inquiry* (Winter 2005–2006), 335-50.

Licasto, Federico, Giuseppina Candore, Domenico Lio, Elisa Porcellini, Giuseppina Colonna-Romano, Claudio Franceschi, and Calogero Caruso. "Innate Immunity and Inflammation in Aging: A Key for Understanding Age-Related Diseases." *Immunity & Aging* (May 18, 2005). Available online at http://www.immunityageing.com/content/2/1/8

Lichtenstein, Alice H., et al. "Diet and Lifestyle Recommendations Revision 2006: A Scientific Statement From the American Heart Association Nutrition Committee." *Circulation* (2006), 82-96. Available online at http://circ.ahajournals.org/cgi/content/short/114/1/82

Loutrari, H., et al. "Mastic Oil from Pistacia lentiscus var. chia Inhibits Growth and Survival of Human K562 Leukemia Cells and Attenuates Angiogenesis." *Nutrition and Cancer* (2006), 86-93.

MacMillan, Amanda. "Will You Live to 100?" *Prevention* (May 2007), 56.

Mariotti, Stefano, Paolo Sansoni, Guiseppe Barbesino, Patrizio Caturegli, Daniela Monti, Andrea Cossarizza, Tamara Giacomelli, Giovanni Passeri, Umberto Fagiolo, Aldo Pinchera, and Claudio Franceschi. "Thyroid and other organ-specific autoantibodies in healthy centenarians." *Lancet* (June 20, 1992), 1506-8.

Martin, Corby K., Stephen D. Anton, Heather Walden, Cheryl Arnett, et al. "Slower eating rate reduces the food intake of men, but not women: Implications for behavioral weight control." *Behaviour Research and Therapy* (October 2007), 2349-59.

Melton, Lisa. "The Antioxidant Myth: A Medical Fairy Tale." *New Scientist* (August 5, 2006), 40-3.

Musick, Marc A., James S. House, and David R. Williams. "Attendance at Religious Services and Mortality in a National Sample." *Journal of Health and Social Behavior* (June 2004), 198-213.

Nagourney, Eric. "A Cardiovascular Argument For Eating Whole Grains." *New York Times* (May 15, 2007) F7. Available online at http://www.nytimes.com/2007/05/15/health/nutrition/15nutr.html

Olshansky, S. Jay, Douglas Passaro, Ronald Hershow, et al. "Peering into the Future of American Longevity." *Discovery Medicine* (April 2005), 130-34.

Olshansky, S. Jay, Leonard Hayflick, and Bruce A Carnes. "Position Statement on Human Aging." *The Journals of Gerontology Series A: Biological Sciences and Medical Sciences* (2002), B292-7.

Perls, Thomas, Iliana V Kohler, Stacy Andersen, Emily Schoenhofen, et al. "Survival of Parents and Siblings of Supercentenarians." *The Journals of Gerontology: Series A: Biological Sciences and Medical Sciences.* (2007), 1028-35.

"Physiology: New studies on goats' milk show it is more beneficial to health than cows' milk." *Obesity, Fitness & Wellness Week* (August 18, 2007), 272.

Poulain, Michel, Gianni Pes, C. Grasland, C. Carru, L. Ferrucci, G. Baggio, C. Franceschi, and L. Deiana. "Identification of a geographic area characterized by extreme longevity in the Sardinia island: the AKEA study." *Experimental Gerontology* (September 2004), 1423-9.

Rowen, Robert Jay. "Artemisinin: From Malaria to Cancer Treatment" *Townsend Letter: The Examiner of Alternative Medicine* (December 2004), 68.

Index

Illustration Credits